Empirical Design in Structural Engineering

Thomas Boothby

Published by Emerald Publishing Limited, Floor 5, Northspring, 21-23 Wellington Street, Leeds LS1 4DL.

ICE Publishing is an imprint of Emerald Publishing Limited

Other ICE Publishing titles:
Empirical Structural Design for Architects, Engineers and Builders
Thomas Boothby. ISBN 978-0-7277-6207-8
Conceptual Structural Design, Third edition: Bridging the Gap Between Architects and Engineers
Olga Popovic Larsen. ISBN 978-0-7277-6598-7
ICE Manual of Structural Design: Buildings
John Bull (Ed). ISBN 978-0-7277-4144-8

A catalogue record for this book is available from the British Library

ISBN 978-0-7277-6633-5

Cover photo: Abstract line structure. shuoshu/Getty Images

Commissioning Editor: Viktoria Hartl-Vida
Assistant Editor: Cathy Sellars
Production Editor: Sirli Manitski

Typeset by Manila Typesetting Company
Index created by LNS Indexing

Empirical Design in Structural Engineering

Contents

To my students and teachers

Acknowledgements

Many friends, colleagues and collaborators have helped me with this book by contributing to the content, by offering suggestions, by offering criticism, or by undertaking to review chapters. These include my friend and colleague Rebecca Napolitano and my friend Donald Friedman. I value Samantha Leonard's knowledgeable professional review of Chapter 8 on forensic engineering. I would also mention contributions by my colleagues Ryan Solnosky and Kevin Parfitt.

I would also like to acknowledge the assistance, advice and guidance I have received from my many collaborators over the years, even on projects not directly related to the subject matter of this book. Within this group, I would mention particularly Elizabeth Smith, Paul Fanning, Sharyn Clough, Dario Coronelli, Giuliana Cardani, Grigor Angjeliu and William Bahnfleth.

Lisa Han and Ian Self, students in the Architectural Engineering Department, were of considerable help in preparing figures.

I gratefully acknowledge the help I have received from the staff of ICE Publishing along the way in producing this book, especially James Hobbs and Viktoria Hartl-Vida, acquisitions editors, and Cathy Sellars and Sirli Manitski, production editors, for assistance with the production of this book.

The dedication is identical to the dedication of *Structural Members and Frames* by Ted Galambos, one of my greatest and best-remembered teachers. In repeating his dedication, I acknowledge my own place in the continuum of teaching and learning.

About the author

Thomas Boothby is Professor of Architectural Engineering at the Pennsylvania State University. He has over 40 years of experience in structural engineering, including 10 years as a structural designer. He has a significant record of publications on the structural interpretation of medieval architecture, the analysis and assessment of masonry and iron bridges, and the application of fibre-reinforced polymers to structural engineering. He is the author of *Engineering Iron and Stone: Understanding Structural Analysis and Design Methods of the Late 19th Century* and *Empirical Design for Architects, Engineers, and Builders*. He is a registered architect and professional engineer and a Fellow and Life Member of the American Society of Civil Engineers.

Preface

This book is a reflection of an outlook I have developed throughout my career, during the 1980s as a design engineer, and later as an academic engineer. I realised at an early stage that an engineer's work involved more than the application of science to problems of construction. An engineer's activity requires the application of experience, which could supplement, replace, or even overrule the results of a rational understanding of structures. In my career as an academic engineer, I spent a large part of my time reviewing the design of unreinforced masonry structures: nineteenth century masonry bridges and gothic or neo-gothic architecture. As structures whose conception was based on experience, these works speak eloquently about the success of empirical design. In fact, much of the contemporary research on these structures attempts to use rational analysis procedures to understand what their builders understood very well empirically. So, while developing an understanding of the successes of empirical design, I started applying empirical design to modern structures and investigating the empirical content of the modern practice of structural engineering. I found that empirical design was a very useful method for teaching engineers and architects to look first at the design of the whole structure and then apply their detailed methods to the elements of the structure. In a previous book, *Empirical Structural Design for Architects, Engineers, and Builders*, I describe systematically how this method can be applied to determining the configuration, materials and sizes of the elements of a structure – the intent of this book was to teach the application of these methods.

Within the past ten years, I have discovered empiricist philosophy and have learned that it can be applied to an understanding of empirical design. The central idea of this school of philosophy is that experience, rather than reason, is the way that we learn about the world. This is also the idea of empiricist design, that design based on prior experience is at least as effective as design based on a scientist's understanding of the world. These investigations have shaped my outlook that structural engineering does not depend wholly on science or reason. Structural engineering is a collection of empirical procedures sometimes informed by science.

Boothby T
ISBN 978-0-7277-6633-5
https://doi.org/10.1680/edse.66335.001

Chapter 1
Empirical design in structural engineering

1.1. Introduction

Empirical design is design by experience that is applied by using past successes directly as a basis for later designs. The experience may be direct personal experience on the part of the designer, or it may be collective experience of previous designers, often distilled into simple rules. Empirical design is an effective means of establishing the initial premises of a design and determining materials, configuration, size and spacing of the components of a structural design. The effectiveness of empirical design is simply that it is using proved procedures to achieve a favourable result in the design of a building structure: a structure that achieves the purpose of resisting applied loads, is an effective complement to the architecture of the building and is of reasonable dimensions.

Empirical design was the only means of completing a building design to the beginning of the nineteenth century and the more effective means during the nineteenth century, so the production of successful buildings predates the appearance of rational design by approximately two millennia. Figure 1.1 illustrates two structures whose design was wholly or partially empirical.

Figure 1.1 View of Amiens, France

Even with the analytical methods available to contemporary engineers, empirical design is commonly used in engineering in two basic circumstances: first, where a product design has become so routine that further detailed design is no longer necessary, and the second where the application of analytical rules is too complicated to apply in routine design.

A house carpenter who sizes rafters, joists, beams and studs in the framing of a house simply by knowledge of what size is required for a given span is practising empirical design. So is an engineer who sizes small lintels in a brick wall with the depth equal to 1/12 of the span, or an engineer who recognises that 90 mm steel angles are suitable as lintels because their legs are approximately the width of a brick and because they can span up to 1800 mm based on a span: depth ratio of 20. Empirical design may be thought to be obsolete in an age in which we are able to calculate the stresses resulting from structural loads and calculate the resistance of the members that resist these loads. However, the remainder of the book will describe the continuing utility of empirical design.

Empirical design is necessary for the establishment of minimum or standard values of the size, materials and configuration of structural elements. The experience with failures of members and connections is often necessary to be able to identify a minimum size or quantity of a structural element. As an example, where the calculated forces in a bolted steel connection may allow the use of 6 mm dia. bolts, a minimum of 12 mm or 14 mm diameter would be observed by most professionals, because a smaller bolt may be subject to failure by unpredicted effects. Similarly, unwritten standards often prevail in the design of other elements of a structure. In the US, 115 mm thick normal weight concrete floor slab is effectively a standard practice, because experience dictates that this is the minimum thickness necessary for fire protection.

Empirical design is useful for the design of elements that are too complex to calculate. Connections furnish many examples of this application of empirical design. It is difficult (and certainly not worth the trouble) to calculate the force acting on a single nail on a plywood sheathing panel subjected to wind loads. The spatial and temporal variations of wind pressure on a cladding panel are very difficult to capture analytically. Similarly, the nails around the perimeter of a sheet of plywood have differing resistances and differing stiffnesses, so that the distribution of forces to the nails is further complicated. The actual resistance of a single nail is also a quantity that is variable depending on the density of the wood, the straightness of the nail and other unknown factors. As a result of these uncertainties, we rely on tables to furnish us with an empirically determined nominal resistance of the nail under defined conditions. Figure 1.2 is an excerpt from the special provisions for design for wind and seismic forces (AWC, 2021) that accompanies the National Design Specification (AWC, 2018) for wood structures in the US, which illustrates the design of wood panel products under the uncertainties of loading and resistance of these connections. Pull-out forces on an anchor bolt in a concrete spread footing furnish a similar example. These are instances of a more general principle that we resort to empirical design when the stresses become too difficult to calculate, or too unpredictable, being based on many other factors.

Empirical design is embedded in modern building codes. As one example, the US steel design code (AISC, 2016) develops rules for the analysis of partially composite steel–concrete beams. For fully composite beams, the rules are a straightforward application of the theory of plasticity

Figure 1.2 Analysis of plywood fastening (AWC, 2021). Courtesy of American Wood Council, Leesburg, VA

Table 4.2A Nominal Unit Shear Capacities for Sheathed Wood-Frame Diaphragms

Blocked Wood Structural Panel Diaphragms[1,2,3,4,6]

Sheathing Grade	Common Nail Size[5] Length (in.) x Shank diameter (in.) x Head diameter (in.)	Minimum Nail Bearing Length in Framing Member or Blocking, l_m (in.)	Minimum Nominal Panel Thickness (in.)	Minimum Nominal Width of Nailed Face at Adjoining Panel Edges and Boundaries (in.)	Nail Spacing (in.) at diaphragm boundaries (all cases), at continuous panel edges parallel to load (Cases 3 & 4), and at all panel edges (Cases 5 & 6)											
					6			4			2-1/2			2		
					Nail Spacing (in.) at other panel edges (Cases 1,2,3, & 4)											
					6			6			4			3		
					v_n (plf)	G_a (kips./in.)		v_n (plf)	G_a (kips./in.)		v_n (plf)	G_a (kips./in.)		v_n (plf)	G_a (kips./in.)	
						OSB	PLY		OSB	PLY		OSB	PLY		OSB	PLY
Structural I	6d (2 x 0.113 x 0.266)	1-1/4	5/16	2	520	15	12	700	8.5	7.5	1050	12	10	1175	20	15
				3	590	12	9.5	785	7.0	6.0	1175	9.5	8.5	1330	17	13
	8d (2-1/2 x 0.131 x 0.281)	1-3/8	3/8	2	755	14	11	1010	9.0	7.5	1485	13	10	1680	21	15
				3	840	12	10	1120	7.5	6.5	1680	10	9.0	1890	18	13
	10d (3 x 0.148 x 0.312)	1-1/2	15/32	2	895	24	17	1190	15	12	1790	20	15	2045	31	21
				3	1010	20	15	1345	12	9.5	2015	16	13	2295	26	18
Sheathing and Single-Floor	6d (2 x 0.113 x 0.266)	1-1/4	5/16	2	475	15	10	630	9.0	7.0	940	13	9.5	1065	21	13
				3	530	12	9.0	700	7.0	6.0	1065	10	8.0	1205	17	12
			3/8	2	520	13	9.5	700	7.0	6.0	1050	10	8.0	1175	18	12
				3	590	10	8.0	785	5.5	5.0	1175	8.5	7.0	1330	14	10
	8d (2-1/2 x 0.131 x 0.281)	1-3/8	3/8	2	670	15	11	895	9.5	7.5	1345	13	9.5	1525	21	13
				3	755	12	9.5	1010	7.5	6.0	1510	11	8.5	1710	18	12
			7/16	2	715	14	10	950	8.5	7.0	1415	12	9.5	1610	20	13
				3	800	11	9.0	1065	7.0	6.0	1595	10	8.0	1805	17	12
			15/32	2	755	13	9.5	1010	7.5	6.5	1485	11	8.5	1680	19	13
				3	840	10	8.5	1120	6.0	5.5	1680	9.0	7.5	1890	15	11
	10d (3 x 0.148 x 0.312)	1-1/2	15/32	2	810	25	15	1080	15	11	1610	21	14	1835	33	18
				3	910	21	14	1205	12	9.5	1820	17	12	2060	28	16
			19/32	2	895	21	14	1190	13	9.5	1790	18	12	2045	28	17
				3	1010	17	12	1345	10	8.0	2015	14	11	2295	24	15

1. Nominal unit shear capacities shall be adjusted in accordance with 4.1.4 to determine ASD allowable unit shear capacity and LRFD factored unit resistance. For general construction requirements see 4.2.7. For specific requirements, see 4.2.8.1 for wood structural panel diaphragms. See Appendix A for common nail dimensions.
2. For species and grades of framing other than Douglas-Fir-Larch or Southern Pine, reduced nominal unit shear capacities shall be determined by multiplying the tabulated nominal unit shear capacity by the Specific Gravity Adjustment Factor = [1-(0.5-G)], where G = Specific Gravity of the framing lumber from the NDS (Table 12.3.3A). The Specific Gravity Adjustment Factor shall not be greater than 1.
3. Apparent shear stiffness values, G_a, are based on nail slip in framing with moisture content less than or equal to 19% at time of fabrication and panel stiffness values for diaphragms constructed with either OSB or 3-plywood panels. When 4-ply, or 5-ply plywood panels or composite panels are used, G_a values shall be permitted to be increased by 1.2.
4. Where moisture content of the framing is greater than 19% at time of fabrication, G_a values shall be multiplied by 0.5.
5. Tabulated nominal unit shear capacities are applicable for carbon steel smooth shank nails of the specified type and size.
6. Diaphragm resistance depends on the direction of continuous adjoining panel edges with respect to the loading direction and direction of framing members, and is independent of the panel orientation.

to flexural members. However, for the calculation of deflections in partially composite beams, the code resorts to the calculation of empirical lower bounds on the deflection.

All the building and material codes that prescribe the design of connections have significant empirical content. Connections in wood structures depend on a variety of fastener capacities that are given in a table, which has been developed based on a limited number of tests. Design values for steel bolts are subject to primarily empirical modifications for the size of the hole, the edge and end distance of the hole and other factors. As an example from the Eurocodes, Eurocode 5 contains a procedure for addressing the highly statically indeterminate distribution of shear forces in a multi-member wood connection, by simply considering statically each potential shear plane (Porteous, 2013, paragraph 8.1.3).

The application of new structural materials develops empirically. Reinforced concrete, when this material first appeared in the nineteenth century, had no analytical procedure for ensuring the strength and safety of a structure built with this material. Early versions of a design procedure for this material relied in part on proportional rules, and in part on a rudimentary, intuitive view of the behaviour of a reinforced concrete structure. This outlook has been described by Boothby and Clough (2017) and will be elaborated in Chapter 3.

Empirical design is an effective tool for completing the initial design of a building. The following cases represent examples of this application of empirical design.

A structural engineer may be required to deliver a preliminary design to an architect for elaboration before they have had the opportunity to size the structure completely or to determine the extent of the reinforcement.

In order to obtain tenders for an accelerated project, an engineer designs the dimensions of the frame in steel or concrete without specifying exactly the sizes, spacing and configuration of the concrete reinforcement. In particular, for a reinforced concrete structure, the sizes of the slabs, joists, beams and columns may be prescribed along with an allowance for the volume of the reinforcement. This allows the project to be estimated and bid, based on dimensions of concrete structure and estimated weight of reinforcement, before the specific structural design is complete. In order to know the appropriate thickness of a floor slab or dimensions of a beam or column, it is necessary to have sufficient experience to be able to choose these sizes effectively.

An architect or structural engineer who determines bay sizes appropriate to a specific structural material or system as an initial step in the structural design process is practising empirical design. Subtleties that result from experience are often part of these decisions. Such subtleties may include modifying the span of the exterior bay or the placement of the exterior columns to reflect differences in loading and resistance, or the use of slightly rectangular bays to reflect the span direction of joists or infill beams and the span of girders.

An arbitrary minimum design value for a bolted connection and an arbitrary minimum depth of a beam-column connection may be specified, again without regard for actual loading conditions. An excerpt from the notes of a structural steel construction document (Penn State

University, 2014) imposes a minimum capacity of a beam connection regardless of the actual load on the beams.

> Design connections using the 'maximum total uniform load' tables in the AISC manual. For non-composite beams, the connection capacity shall be at least 50% of the maximum total uniform load, for composite beams, the connection capacity shall be at least 80% of the maximum total uniform load.

In contemporary engineering practice, the successful application of optimisation to engineering design problems requires the insertion of a credible initial estimate, known as the seed, to ensure that the optimisation routine converges on a practical solution. Most ordinary design techniques are iterative, but iterative processes are more expedient if there is a reasonable starting point. Empirical design can provide a suitable initial design to be used as a starting point for optimisation.

Further and more detailed examples of the use of empirical design are discussed throughout the remainder of this book both from a general point of view and in specific applications.

Empirical design has a basis in epistemology, the philosophy of the acquisition and verification of knowledge, that lends credit to its effectiveness as a design method. Empiricism is the outlook that knowledge is assembled from experience or observation without recourse to reason, and that the application of reason does not provide useful information about the universe. A rationalist outlook is that by reasoning, even in the absence of concrete evidence, we can make significant discoveries about the functioning of materials or of natural processes. Although empiricists understand mathematics and find it interesting, they doubt that mathematical manipulations tell us anything useful. An example of this, which will be discussed later, is the Euler formula for the buckling of an elastic column. This is an elegant mathematical theory that some empirically minded engineers have dismissed as not useful for evaluating the columns in real buildings because these columns are not slender enough for the formula to be applicable. In fact, even the building codes that acknowledge the merit of this formula use an empirically determined means of constructing the curve within the range of columns of practical interest.

An examination of the history of building will show that concepts from natural science were not directly applied to the understanding of building systems until the nineteenth century. Up to that time, the sizes of structural elements were based on experience, usually transmitted through some sort of proportional rules. The creation of usable, enduring structures by ancient building methods, which were exclusively empirical, speaks to the potential success of empirical design in similarly creating usable and enduring structures. On the other hand, a very modern outlook on engineering, the idea of data-driven design, or data-driven artificial intelligence, uses empirical information to achieve a design result without recourse to a physics-based model.

The application of empirical design in practice requires the consideration of the ethics of using a method only indirectly endorsed by the building profession. The ethical dimensions of empirical design are considered in a later chapter. The fundamental ethical question is to what extent an engineer may choose to base their decisions on empirical engineering. This question is closely connected to the discussion of building codes and empirical design. The premise of most of the building codes is that design of structures should follow rational principles. However, there are occasions when rational design is simply not possible. In cases like this, it may be necessary to consider experience and to act accordingly. There are other cases in which resorting to experience and trusting calculations are two alternative pathways to arriving at a competent design. In these cases, there is considerable pressure on an engineer to follow the practice of presenting calculations according to the requirements of a building code. There is merit to completing an empirical design in parallel, and using the two designs, empirical and rational, as a means of checking the other. In the end, it may be argued it is ethical to use a design method that has been proved over 20 centuries.

The following is a chapter-by-chapter summary of the remainder of the book.

Chapter 2 describes the philosophical basis of empirical design, primarily from the point of view of epistemology, the branch of philosophy that deals with the acquisition and validation of knowledge.

Chapter 3 outlines a more practical view of empiricist philosophy, relating it to works on the philosophy of engineering.

Chapter 4 examines the methods used by empirical builders throughout the past two millennia and these builders' successes that can be attributed to empirical design.

Chapter 5 discusses how contemporary building codes and contemporary engineering practice may be more conventionalist and empiricist and less rational.

Chapter 6 examines the ethical issues in using empirical design in contemporary engineering practice.

Chapters 7–10 are case studies in the application of empirical design, from the point of view of a twenty-first century engineer.

Chapter 7: preservation engineering
Chapter 8: forensic engineering
Chapter 9: wood-framed building superstructure
Chapter 10: foundation design

Chapter 11 looks at further potential uses of engineering design, including the use of empirical design in artificial intelligence and optimisation applications.

Chapter 12 presents conclusions based on the information from the previous chapters.

1.2. Conclusions

The premise of the following book, then, is that design by experience has a long history of effective application to the design of buildings and can continue to be used, consciously or unconsciously, by contemporary engineers. Earlier chapters will be an investigation into empiricism and the engineering implications of empiricism, the philosophical basis of empirical design. These investigations will be undertaken to understand how an understanding of the processes of nature can lead engineers to understanding the authority of experience. The following chapters will describe how contemporary building codes have embedded empirical content, through an analysis of portions of contemporary building codes, such as the Eurocodes and the ASCE 7 code for structural loading in the USA.

The ways in which empirical design can be effectively used are in the assessment of historic structures, the design of foundations, the completion of routine or repetitive designs, or in the application to structures whose analysis is overwhelmingly complex. In the latter case, empirical design may be used in its traditional form or may be implemented by computer methods, including data-driven methods or artificial intelligence.

Empirical design has further value as a way of determining approximate configurations, materials and sizes of structural elements for buildings. There are situations in which this is a necessity for practising engineers, such as fast-track construction, or for developing documents for tenders in a reinforced concrete building when the exact design of the reinforcing steel is not available.

REFERENCES

AISC (American Institute of Steel Construction) (2016) AISC 360-16: Specification for structural steel buildings. AISC, Chicago, IL, USA.

AWC (American Wood Council) (2018) *National Design Specification*. AWC, Leesburg, VA, USA.

AWC (American Wood Council) (2021) *Special Design Provisions for Wind and Seismic (SDPWS)*. AWC, Leesburg, VA, USA.

Boothby TE and Clough S (2017) Empiricist and rationalist approaches to the design of concrete structures. *APT Bulletin* **48(1)**: 6–14.

Penn State University (2014) *Construction Documents for the Chemical/biological Engineering Building, Sheet S001. General Structural Notes*. Pennsylvania State University, PA, USA.

Porteous J (2013) *Designer's Guide to Eurocode 5*. ICE Publishing, London, UK.

Boothby T
ISBN 978-0-7277-6633-5
https://doi.org/10.1680/edse.66335.009

Chapter 2
Philosophical empiricism

2.1. Introduction

In this chapter, some ideas are introduced from enlightenment and contemporary philosophers about epistemology, 'the branch of philosophy which investigates origins, structure, methods, and validity of knowledge' (Runes, 1983). This discussion is intended to demonstrate that it is philosophically defensible to rely on empirical evidence alone. An engineer persuaded by empirical evidence of the validity of a given structure may not need to be further persuaded by taking a rational or analytical approach to investigating the structure. Clearly, a structure can often be accepted on the basis of both rational and empirical evidence. Those cases in which there may be contradictions between the two methods are discussed in Chapter 6, where the ethical aspects of the use of empirical design are investigated.

2.2. Philosophical considerations

To understand this discussion of empirical design in a philosophical context, it is necessary to define philosophical ideas including scientific realism, anti-realism, empiricism and rationalism. According to the Stanford Encyclopedia of Philosophy (2022)

> Scientific realism: Scientific realism is a positive epistemic attitude toward the content of our best theories and models, recommending belief in both observable and unobservable aspects of the world described by the sciences.
>
> From the article 'Empiricism and Scientific Realism': In the historical development of realism, arguably the most important strains of antirealism have been varieties of empiricism which, given their emphasis on experience as a source and subject matter of knowledge, are naturally set against the idea of knowledge of unobservables.

Further, from Runes (1983)

> Realism means… in epistemology: that sense experience reports a true and uninterrupted, if limited, account of objects: that it is possible to have faithful and direct knowledge of the actual world.

In addition to the tension between realism and empiricism described above, there is a fundamental difference in outlook between empiricism and rationalism. In brief, rationalists believe that reason is the primary source of knowledge about the world and that empirical observation is a source of error, while empiricists believe that experience is a more important source of knowledge than reason. This passage, from the 'Empiricism vs. rationalism' article, captures some of the subtleties of a widespread and centuries-long debate.

> To be a rationalist, however, does not require one to claim that our knowledge is acquired independently of any experience: at its core, the Cartesian cogito depends on our reflective, intuitive awareness of the existence of occurrent thought. Rationalists generally develop their view in two steps. First, they argue that there are cases where the content of our concepts or knowledge outstrips the information that sense experience can provide. Second, they construct accounts of how reason, in some form or other, provides that additional information about the external world.
>
> Empiricism: Most empiricists present complementary lines of thought [to the above discussion of rationalism]. First, they develop accounts of how experience alone – sense experience, reflective experience, or a combination of the two – provides the information that rationalists cite, insofar as we have it in the first place. Second, while empiricists attack the rationalists' accounts of how reason is a primary source of concepts or knowledge, they show that reflective understanding can and usually does supply some of the missing links.

A further idea in the philosophy of science, conventionalism, holds that (Runes, 1983)

> Any doctrine according to which *a priori* truth, or the truth of propositions of logic, or the truth of propositions (or of sentences) demonstrable by purely logical means, is a matter of linguistic or postulational convention (and thus not absolute in character).

According to the conventionalist view, the truth of an engineering finding, such as the calculation of the deflection of a beam, is not absolute, but is accepted by convention as a means of making further progress in understanding. As conventionalism is most important for understanding the relation of building codes to engineering design, the discussion of conventionalism will be reserved for Chapter 5.

The philosophical ideas described above: realism, anti-realism, empiricism, conventionalism and rationalism have an impact on structural engineering and the practice of structural engineering specifically in the way that the behaviour of artefacts in the natural world is to be understood by an engineer. The primary occupation of the engineer is the manipulation of human-made artefacts to serve engineering purposes. To place a wooden beam over a door opening to support the wall above, the engineer needs an understanding of what size the beam should be, where its supports should be, how large its supports should be and how long to expect that the beam will function. All these decisions are mediated by the effect of nature on the beam and on the objects that it is supporting. It is of some importance to understand whether the rules used to design the beam have some fundamental explanation in mechanics, whether the explanation is simply developed phenomenologically, based on observed phenomena, or whether the beam can be designed by experience, without resorting to an understanding of natural laws. We will outline the applicability of these ideas to the investigations necessary to complete tasks in structural engineering.

A further meaning of the term empiricism in Runes' dictionary is especially applicable to engineering design activity. 'A practice method, or methodology based on direct observation or on immediate experience, or which precludes analysis or reflection, or which employs experimentation or systematic induction as opposed to discursive, deductive, speculative, transcendental, or dialectic procedures.' The definitions of empiricism are weighted towards knowledge acquisition by experience and away from deductive reasoning as a means of knowledge acquisition. The above excerpt from Runes is more closely related to practice and could be considered as a summary of empirical design.

In philosophy, empiricism is distinct from rationalism, which, in its most developed form, considers knowledge acquired by the senses as erroneous and misleading and only recognises as truthful knowledge that is arrived at through the application of reason. Rationalism, again according to Runes, is 'A theory of philosophy in which the criterion of truth is not sensory but intellectual and deductive.' Moreover, rationalism is 'usually associated with an attempt to introduce mathematical methods into philosophy'.

Rationalism is a concept that has been very generally applied to the study of the natural sciences beyond mathematics, in which deductive reasoning, often through mathematical models, is employed to solve problems or to develop theories. Of particular concern to this work are theories of mechanics, in which a rationalist outlook is usually applied. The term 'rational mechanics' was introduced into engineering education and applied to the development of statics, dynamics and deformations according to mathematical rules. Many texts on rational mechanics use an axiomatic treatment of mechanics, reducing forces, displacements, velocities and so on to mathematical objects, without the necessity of a connection to measurable quantities. Truesdell (1991), for example, explains rational mechanics solely in symbolic mathematical form, without diagrams or concrete examples. Related to rational mechanics is the concept of 'applied mechanics', in which mechanical concepts (force, displacement, velocity, etc.) are applied directly to objects of interest: reactions and bending moments in a beam, forces in a truss system, simple harmonic motion and so on. Although applied mechanics is also developed in mathematical terms, the mathematical treatment is more ad hoc than axiomatic: the mathematical concepts are developed separately for the solution of individual types of problems, hence calculus is separately applied to the determination of centre of gravity and to the calculation of deflections in a beam.

The distinction between the application of rational mechanical principles or empirical rules to the design of a structure can be related to engineering practice through a customary way of thinking about engineering. If one accepts that natural science explains the phenomena of the natural world, and that an engineer can estimate these phenomena through the same logical pathways as the scientist and base their designs on this knowledge, then the object that they have designed is a product of reason. This outlook may be called 'rational engineering' or 'analytical' in opposition to 'empirical' or 'experience-based'.

Table 2.1 shows the distinction between rational/analytical design and empirical design in the case of the design of a column. The mathematics of buckling of compression members is highly developed. As a result, it is possible to calculate a critical buckling load for any steel column. Doubt may appear in that in a column of practical interest, the buckling equations are misleading regarding the strength of the columns. A steel pipe column 200 mm in diameter and 4000 mm long has a slenderness ratio of approximately 40. This is a very typical proportion for a steel column. However, the type of elastic buckling predicted by the rational equations for

Table 2.1 A comparison of the empirical or rational basis for the design of a column

	Rational	Empirical
Governing formula	Euler buckling formula	Length/diameter ratio
How arrived at	Mathematics applied to idealised object	Experience with favourable length/diameter ratios Pipe column wall thickness estimated by experience
Controlling factors	Cross-sectional area, diameter, slenderness	Diameter of column
Shortcomings	Intermediate length columns require application of phenomenological or empirical rule	Indefinite in determination of wall thickness

column strength does not occur at slenderness ratios this low. So, while an empiricist may maintain that the formulas that we use in design are untrustworthy, as the buckling formula is misleading concerning the behaviour of intermediate length columns, the rationalist would say that the buckling equation can be corrected by a rational consideration of the out-of-straightness of the column, eccentricity of load application on the column and residual stresses. The variations in support conditions also have a very large influence on the buckling behaviour and the strength of a column and also require empiricist or rationalist correction. This point is discussed further in Chapters 3 and 5.

There is an implicit presumption of rationality in engineering textbooks. Wight (2016) starts from general principles, listings of design assumptions and descriptions of the mechanics of reinforced concrete reduced to procedures for estimating curvature, ultimate moment and so on. From the introduction to Nelson and McCormac (2002)

> Structural analysis as we know it today evolved over several thousand years. During this time, many types of structures such as beams arches, trusses, and frames were used in construction for hundreds and even thousands of years before satisfactory methods of analysis were developed for them. Though ancient engineers showed some understanding of structural behavior (as evidenced by their construction of great bridges, sailing vessels, cathedrals, etc.) real progress with structural analysis occurred only in the last 175 years.

According to Ambrose and Tripeny (2016), 'most of engineering design and investigation is based on applications of the science of mechanics'. Other engineering texts present similar generalisations. Heyman (1999) has a narrative of scientific design in the eighteenth century overcoming the tradition-bound empirical builders.

As implied by the quotation from Nelson and McCormac above, a fully rationalist approach is not necessary to conceive effective structures, as will be shown in the following chapters.

This point reinforces the importance of the design and construction of buildings using a non-rationalist approach. The rational examination of structural behaviour applied to the predicted behaviour of the designed object is not the only set of principles that can be used in engineering design. As will be demonstrated in Chapter 4, there is a long history of building without rational methods, in which effective structures were created by cautious experimentation within a tradition of building knowledge. Even when a rational approach is invoked, there are underlying empirical principles that are required to justify the selection of a particular rational method, and the rational method nearly always requires empirical correction, a point that will be discussed throughout this book.

In the face of this evidence, it is problematic to accept categorically that engineering is a rationalist endeavour. Cartwright (1983) presents evidence concerning natural science itself, on which rational engineering depends. According to Cartwright, rather than being a unified pursuit, natural science contains a variety of separate theories and requires constant phenomenological or empiricist correction. Engineering is farther removed than natural science from measurable phenomena observing identifiable laws in a consistent manner. The nature of engineering makes such an ordered presentation impossible. Instead, structural engineering can be pursued as a strictly empirical activity, as shown in Chapter 4. In Chapter 3, the nature of the empiricist corrections applied to rationalist design procedures are presented. Although the findings of rationalist engineering and empiricist design are often similar, the cases in which there are contradictions are discussed in Chapter 6, which concerns the ethics of the application of empirical design.

2.3. Conclusions

Epistemology, as applied to the study of natural sciences, can be extended to engineering with reservations that the purposes of the two pursuits are different: natural science is explanatory, while engineering is concerned with the construction of artefacts. The tension between rationalism and empiricism that is present in the natural sciences is also present in engineering, while the tension between realism and anti-realism is also apparent. The basis of empirical design is empiricism: the conviction that the knowledge of nature and the estimation of the actions of artefacts is primarily discovered by experience.

REFERENCES

Ambrose J and Tripeny P (2016) *Simplified Engineering for Architects and Builders*, 12th edn. Wiley, New York, NY, USA.

Cartwright N (1983) *How the Laws of Physics Lie*. Oxford University Press, Oxford, UK.

Heyman J (1999) *The Science of Structural Engineering*. Imperial College Press, London, UK.

Nelson JK and McCormac JC (2002) *Structural Analysis: Using Classical and Matrix Methods*. Wiley, New York, NY, USA.

Runes D (1983) *Dictionary of Philosophy*. Philosophical Library, New York, NY, USA.

Stanford Encyclopedia of Philosophy (2022) https://plato.stanford.edu/ (accessed 29/07/2022).

Truesdell CA (1991) *A First Course in Continuum Rational Mechanics*. Academic Press, Boston, MA, USA.

Wight J (2016) *Reinforced Concrete: Mechanics and Design*, 7th edn. Pearson, Englewood Cliffs, NJ, USA.

Boothby T
ISBN 978-0-7277-6633-5
https://doi.org/10.1680/edse.66335.015

Chapter 3
Engineering empiricism

3.1. Recapitulation of philosophical empiricism

In the previous chapter, it was noted that, although engineering may appear to be a rational pursuit, founded on the a priori development of models of behaviour and design, the practice of engineering incorporates significant elements of empiricism. Similarly, although the outlook of most engineers is realist – that is, the principles that an engineer works with are thought to be related to enduring and identifiable laws of nature – much of the practice of engineering is ad hoc, contingent and anti-realist. These ideas will be elaborated further in this chapter, with a particular emphasis on how the activities of an engineer fit into the epistemological concepts that we have identified. Following this description of the philosophical ideas underlying the practice of engineering, we will identify and discuss the areas in which empiricism is used in structural engineering, either on its own or as a supplement to rational analysis.

In this chapter, writings are introduced on applied science and engineering that describe the distinction between a realist and non-realist viewpoint or a rationalist or empiricist viewpoint – realism and rationalism are different, widely held views among scientists. Realists assert the existence of natural laws as a real, unchangeable object. Some aspects of the non-realist viewpoint, exemplified by Cartwright (1983) will be described in this chapter, as this outlook has significant implications for engineering. Of further importance is the difference between a rationalist and an empiricist viewpoint, in that an empiricist supposes that knowledge comes primarily from experience. In the application to engineering, an empiricist may mistrust ideas that come from rational analysis while embracing viewpoints about structure design that result from experience.

In the previous chapter, we have briefly mentioned Cartwright's critical discussion of the realist view of science, along with her proposal for a less organised, more Aristotelean outlook on science and the investigation of science. Her comments frequently apply also to engineering. She describes notions of overlapping physical laws that do not individually address all phenomena, the corrections that need to be made to statements of natural laws and the extreme philosophical difficulties in the composition of effects.

A further treatment of conventionalism, the epistemological outlook that knowledge is based on agreed-on conventions rather than an interpretation of natural laws, is reserved for Chapter 5.

3.2. Relationship of empiricism to engineering activity

The philosophical discussions of empiricism and rationalism, which apply natural science, can also be related to engineering activity to distinguish between scientific research and engineering design. Engineering design is a very different activity from scientific investigation, although both make use of the collection of facts from nature and the development of explanatory theories for these facts. The purpose of engineering activity is constructive – that is, to develop a useful artefact. In pursuit of this objective, the engineer takes a different attitude from the natural scientist.

Staples (2014, 2015) discusses the epistemological differences between science and engineering. Staples identifies the characteristics of engineering theories that make them very different from scientific theories, namely, that they are narrower in scope, may have internal overlaps and contradictions, that they are directed towards the creation of an artefact rather than explaining any phenomena. The author further identifies features of engineering theories and engineering procedures that are more directly connected to empiricism or empirical design: the use of data tables without the support of a theory, for instance. The fact that an engineer is primarily concerned with obtaining a result allows the adoption of stances contrary to theory or makes the engineer more willing to use conservative approximations.

A critique of realism is necessary in a critical review of engineering to understand the absence of real, unchangeable laws of nature, at least as they apply to engineering activity. In place of a single covering law, an engineer substitutes multiple overlapping theories, empiricist corrections, empiricist appeals and appeals to practice. A simple instance of the engineering profession's willingness to use multiple overlapping theories: the explanation of bending of a beam requires a different approach for different materials such as concrete, wood or steel. Being engaged in an activity with the objective of the creation of artefacts, engineers are comfortable using a variety of means to achieve this end. Although they may understand results from science: from mechanics, thermodynamics, chemistry and so on, they do not necessarily invoke these results in their daily activity. Instead, they often rely on previous experiences: repeating previous successes and avoiding previous failures, the experiences of others, including colleagues, technicians and sales engineers. As described in this chapter, engineers often resort to the use of data for which they do not necessarily understand the source or the underlying ideas. The theories that they do apply are usually applicable to specific materials or specific assemblies rather than being theories that cover a wide variety of situations. As will be discussed in Chapter 5, the codes that engineers use often have an empirical basis and certainly do not follow a consistent, logical, comprehensive approach to material strength. All these practices of the engineering profession can be described as empiricist.

While scientists are more interested in a theory of wider scope, an engineer is willing to accept multiple overlapping theories, narrower in scope with more precision. An example of such a collection of theories could be found in the investigations of stability in steel and in concrete rigid frame buildings according to the US codes AISC 360 (AISC, 2016) for steel and ACI 318-19 (ACI, 2019) for reinforced concrete. The two theories cover the same phenomenon, the reduced stability of a column in a rigid frame when the columns are subjected to axial compression. In both cases, the method for stability analysis is a direct application of moment magnifiers. This calculation of moment magnifiers in the concrete code is similar to the direct analysis method presented in Appendix 8 of AISC 360 in the steel code but differs in the determination of effective lengths of the columns, in the omission of a magnifier for braced frame forces and in other features. Similarly, Eurocode 3 contains moment magnifiers to account for steel columns (5.5.4(1)) and frames (5.2.6.2(3)), while a similar treatment for concrete structures using different formulas can also be found in Eurocode 2 (5.8.7.3) (BSI, 2004, 2005). In steel design, alternative methods may be applied, such as the effective length method, or the use of notional loading. Similarly, significant overlap exists between the theories adopted for the determination of column strength between wood and steel design. The design of steel compression members is covered by AISC 360, while the design of wood compression members is covered by the *National Design Specification* (AWC, 2018). The calculation of the buckling load is primarily used to scale the empirical interpolation formula used for the determination of the maximum load on columns of practical interest. Although the interpolation formula has different expressions for the wood and steel columns, their graphs have similar shapes. The factors that make the columns deviate from ideal Euler buckling are quite different for wood and steel: in steel, primarily residual stresses and partially out of straightness; in wood, non-uniformity and grade-permitted defects of knots, slope of grain and warping (AWC, 2018). No attempts have been made to unify the two theories because there is no practical necessity to do so.

The differences between engineering and scientific investigation imply an empiricist outlook on the part of the design engineer. An engineer is prepared to accept facts, sometimes without a rational justification, justifying the place of empiricism within engineering in general. The lack of concern about the use of multiple theories or the use of data tables without understanding where the data come from are both cited by Staples (2014) as engineering procedures. Both are manifestations of empiricism. Cartwright (1983) notes 'in contrast that engineering focuses on descriptive adequacy at the expense of explanatory power'. This is a similar statement to Staples and notes that in the engineer's interest of achieving a result expediently, the engineer is willing to sacrifice explanatory power. Moreover, facts determined rationally, even those that are followed by scientists, require empiricist corrections. These corrections are described by Cartwright.

> If the fundamental laws are true, they should give a correct account of what happens when they are applied in specific circumstances. But they do not. If we follow out their consequences, we generally find that the fundamental laws go wrong; they are put right by the judicious corrections of the applied physicist or the research engineer.

In engineering, these corrections may be classified as empiricist appeals. An empiricist appeal is a justification of an uncertain rational design procedure by empirical observation (Boothby and Clough, 2017). Many of the procedures that are followed in engineering have only limited rational justification. To be useful, an explicit or implied appeal to empiricism may be required: the most common form of an empiricist appeal is that a procedure, however uncertain rationally, has been found to be effective.

An engineering model should be no more complex than it needs to be. It should not be so complicated that it cannot be understood, nor should it require the input of unknown parameters. The result of this is that a successful analytical model in engineering is a summary of a natural phenomenon, neither a concrete entity, nor a fully formed natural law. According to Staples (2014) 'Engineers… are often satisfied with theories that merely predict phenomena, even if they have no explanatory power or ontological correspondence with the world.' It is an error to ascribe any sort of realism to such a model – the model exists only for the purpose of explaining (partially) a phenomenon that can result in a design. Nevertheless, it is often the case that such explanatory models are extended into real objects. For instance, a column base, as in Figure 3.1, may be modelled as a hinge, but it is an error to suppose on this account that no moment is transferred. The model 'the hinge' used in explaining the column is not a concrete phenomenon, only a modelling assumption, and it is necessary to consider the likely actual conditions in proceeding further with the design. The 'buckling load' on a column is not the load that causes instability in a real column, but a quantity that engineers use for assessing the relative stability of a column.

3.3. Trial and error

Staples and other authors describe the correction of errors as part of the progress of engineering. He identifies the label of 'trial and error' as a 'pejorative misconception of engineering'. Surely, he says, the trials are there, and surely errors are noted, but trials that follow an error are not random, but are directed towards improving the previous trial on intelligent grounds. This statement could also be applied to empirical design in engineering. The term 'trial and error' is applied in an equally pejorative sense to pre-rational engineering (Mora Alonso-Muñoyerro *et al.*, 2014; Romel *et al.*, 2020) – in ancient times, as in contemporary engineering, the trials were justified by previous experience and the correction of the errors was targeted to the apparent cause of the errors. Moreover, the use of 'trial and error' is identified as an engineering myth by Koen (2013). A more credible account of the way that empirical design proceeds – in this instance the rapid development of aseismic forms in traditional architecture – is given in Jorquera *et al.* (2017). Ancient builders surely noted errors in the form of cracks, settlement and, rarely, collapse, but their following trial was a deliberate effort to mitigate the conditions that caused the error in the previous case. As a more up-to-date

Figure 3.1 Typical steel column base. Although this base is modelled as a hinge, there is considerable bending capacity due to the spread of the anchor bolts and the stiffness of the base plate

example, unpredicted failures of building structures in earthquakes led to directed procedures for better earthquake-resistant construction, or to repair some of the previous errors. Steel beams with the flanges welded directly to the column flanges, using full penetration groove welds, create lateral force-resistant frames of a type widely used for earthquake resistance. This type of connection is now known as pre-Northridge connection as many of these connections endured brittle fracture in the 1994 Northridge California earthquake (Bonowitz and Maison, 2003; Lee and Foutch, 2002; Miller, 1998) although there were no building collapses and no loss of life associated with these failures. The modifications to this connection type were developed specifically to remove the conditions that gave rise to the failure of the welded connections: to ensure that the beams were weaker than the columns and that the connection between beams and columns had sufficient ductility.

3.4. The uses of empiricism

In this section, we outline the application of empirical design to engineering activity – in which cases of this activity take place in contemporary structural design or in which cases design methods might profit from the adaptation of an empirical outlook. There are two primary situations in which empirical design is useful as a contemporary design activity: when an element of a structure is produced routinely according to customary or standard practices, or when a structure is too complicated to be analysed effectively. In the second case, there is often recourse to a semi-empirical method – that is, a method that relies on empirical evidence to supplement the results of a rational analysis (Hayes, 1915). The empirical design method is also particularly useful for the preliminary sizing of structural elements and the adoption of conventional solutions for building structures at the preliminary design stage. Experienced engineers do this without really thinking about the rules, by relying on their previous experience with a particular building type.

Additional uses of empirical design include the assessment of existing structures, the development of new materials and methods for which an empirically proved rational design method does not exist. A more modern use of empirical design applies to automatic optimisation routines and to parametric design. In optimisation, it is necessary to start with a feasible design, called a 'seed' in the optimisation process. Such a conventional design can be produced quickly using empirical design. In data-driven analysis, the physics behind a phenomenon may be overlooked in favour of simply collecting and correlating input and output data.

3.5. Simplifying complex structures

Empirical design is applicable to cases in which the analysis of a structure is excessively complicated or has too many uncertainties for a rational result to be easily obtainable or to make sense in construction. This occurs frequently in engineering practice. Often, the lesser importance of a feature and the difficulties of an exact analysis dictate that the feature be designed empirically, or at least that an empiricist appeal is necessary.

Light-gauge cold-formed steel often fits the description of structures for which it is not effective to calculate by analytical methods. The conditions of local buckling of the elements of this type of structure make the exact determination of stress limits very difficult. For some applications – Z purlins in roof structures, for instance – there are other confounding variables of continuity, non-rigid support, combined torsion and bending, asymmetric buckling and so on that render precise calculation for the actual conditions impossible. For Z purlins with a floating roof system (Butler, 2023), for example, there is no continuous bracing of the compression flange of a light-gauge Z purlin, so any buckling analysis of the purlins has to take account of the intermittent elastic support at bridging points. Figure 3.2 shows the anchorage of the roof in this system, which allows both sliding of the roof due to thermal expansion and sliding of the supporting purlin.

Cases of this type can be addressed effectively only by empiricist appeal. In this case, the member may be designed based on the bending stress based on Z-section properties, possibly reduced due to the reduced effective area of the Z purlin in bending, omitting the torsional conditions, conditions of elastic restraint and other issues. Although the omission of torsion from these calculations may be perceived as over-simplifying, this method of calculating incorporates an empiricist appeal: that structures have been designed using this procedure for decades and that it is unnecessary to adopt a more elaborate analysis or design method.

In the analysis of wind and seismic forces in a building structure, a roof system of steel joists supporting a sheet metal roof deck is considered semi-rigid. An exact analysis of a semi-rigid diaphragm is needlessly complicated and certainly not worth the effort of an extended analysis. In a seminar produced by a structural analysis software vendor, the disadvantages of modelling semi-rigid diaphragms are pointed out. It is necessary to know the shear properties of the steel deck interacting with the open-web steel joists, and it is necessary to understand the support conditions of the diaphragm and the details of force transfer between the lateral force resisting system and the diaphragm (Structural Engineering Solutions, 2022). All this information is applied primarily to the distribution of storey forces into the lateral force-resisting system. It is not explained how to develop connections between the steel deck and its supporting structure that resist the required shear forces (this is left to the diaphragm design manual (SDI, 2004)). The result is that it is really not possible to know or understand sufficiently the information that

Figure 3.2 Floating corrugated metal roof system. This system is meant to arrest the development of roof leaks by having a continuous, uninterrupted roof surface. The clips that support the roof over the purlins are allowed to move perpendicularly to the purlins. This reduces the effectiveness of the bracing of the roof purlins. Courtesy of Butler Manufacturing

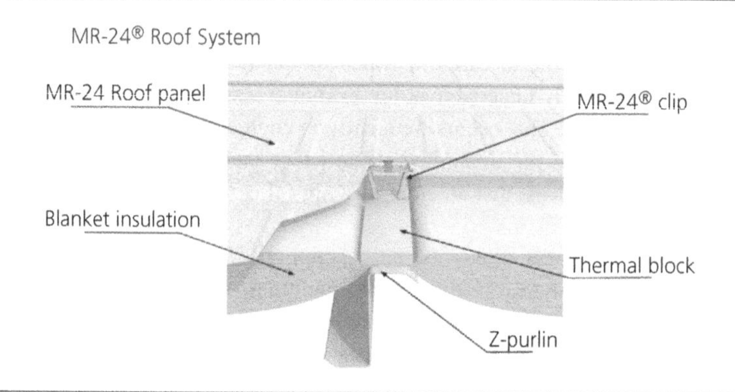

is needed to determine the distribution of forces to a semi-rigid diaphragm, or how to design the connections between the diaphragm and the lateral force-resisting system (shear walls or braced frames) or to design the diaphragm itself. There are two widely accepted means of overcoming these difficulties. The first is to model either a fully rigid or fully flexible diaphragm. The second is to adopt empirical rules, such as those in the Steel Deck Institute (SDI, 2004) manual for the design of the system of connection between a steel deck and supporting structure.

The designer of a bolted connection, for instance, refers to the theories and the limited test data around the failure of bolts in a connection, and then refers to the design rules formulated for the design of a body of this sort. The designer is not necessarily equipped to connect the theory and the design rules, and the theory, or properly the multiple overlapping theories, have elements of uncertainty. As the design rules and data do not cover every situation, the designer's ability to design complex connections depends in part on the ability to see beyond all the test data and all the specification rules and focus on something that experience shows is likely to work.

Two-way reinforced concrete slabs are another extremely complex type of structure encountered in everyday practice. In the design of these systems, a designer has two ways of approaching complexity. The first is to rely on the simplifications used from the 1960s to about 2010, such as the equivalent frame method, or the direct design, which reduces the two-way problem to a more familiar type of one-way system, with modifications made for some of the complexities of behaviour. These two procedures yield very different design results and, again, a designer must use their own experience to resolve the differences between the results of the two different analyses. Neither of these procedures, furthermore, give reliable information about the behaviour of the system at the edges, at re-entrant corners, or at openings in unfavourable locations in the slab. In contemporary practice, it is also possible to use a computer programme to model the entire slab and develop a complete summary of the bending moments and shears in both directions. This procedure has the opposite result: too much information, seemingly too precise, about the response of the two-way system (Figure 3.3). The result does not come in a form that is suitable for design, consisting of contours of internal forces, without any regularity. The results are also sensitive to boundary conditions, properties of the slab and other factors, and do not take account of force redistribution due to cracking of the concrete. In order to mediate between the computer results and the results of the traditional hand method, the designer is forced to rely on their own experience or the experience of others to interpret the results.

In most existing bridge or building applications, masonry arches were empirically designed, so that the evaluation of a masonry arch merits an empirical analysis. In addition to the difficulties of analysing a masonry arch, a masonry arch system presents further challenges. In a bridge, the arch takes the form of a barrel vault provided with rigid spandrel walls on each side, haunching – stones placed on the back of the arch in unknown extent – and fill. The analysis of the interactions between unit and mortar, arch barrel and spandrel walls, arch barrel and haunching, barrel, haunching and fill are complex, involving solid–solid contact, behaviour of granular materials in the fill, the presence of cracking and other confounding factors, so that an analytical treatment of such a structure is problematic. In many cases, the simple application of empirical rules (such as the rules of Rankine (1865) or Trautwine (1874)) and visual inspection

Figure 3.3 Typical bay, two-way system design contours. This is a sample of the contours of the maximum bending moment in one direction throughout the slab. If this material were to be used for design, it would be simplified (Figure prepared by Ian Self)

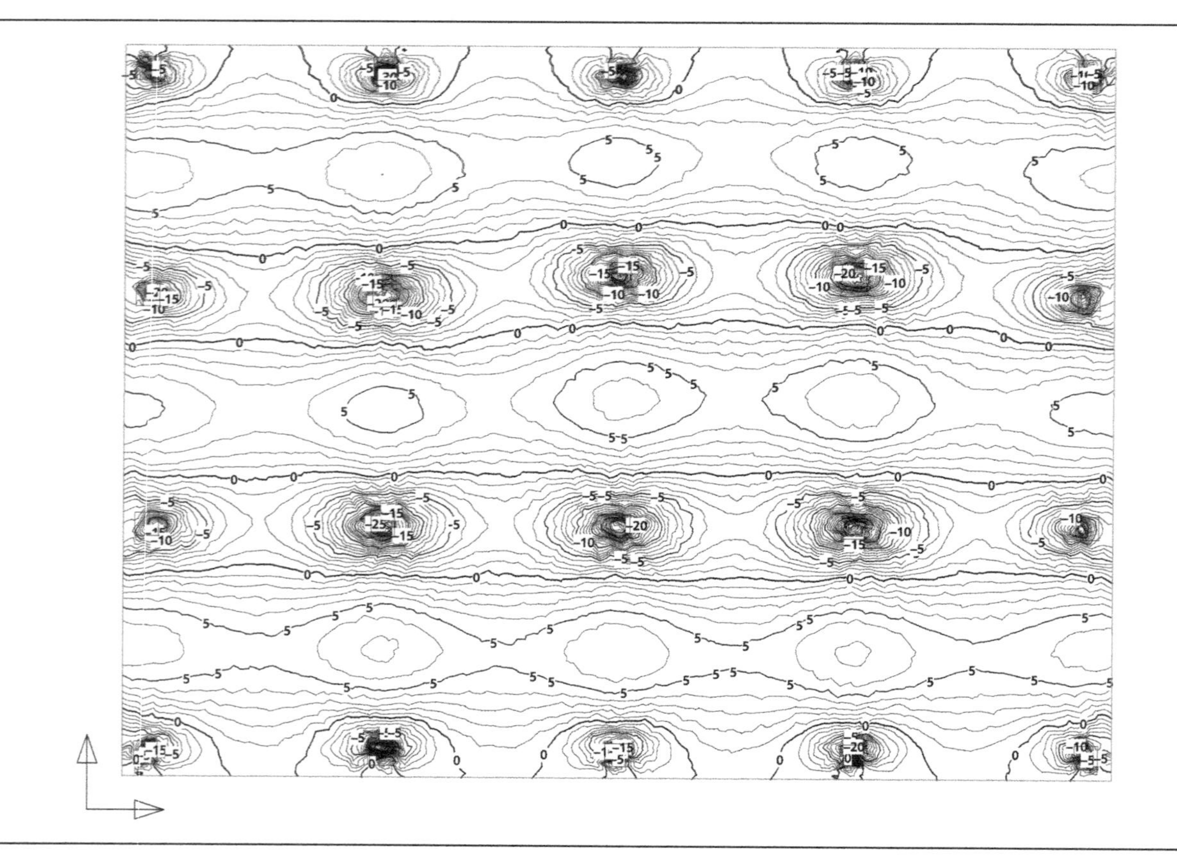

furnish information that is more reliable than an extended analysis using uncertain parameters. This point was previously made by the author (Boothby, 2020).

3.6. Adoption of new materials

Empirical design is often used during the development of new materials, new processes, or new assemblies. In the following examples, two fundamental features will also be highlighted of the relationship between empiricist and rationalist design that are common to many other examples. The first is that new construction techniques are introduced as empirical procedures and, as the practice becomes more widespread, more elaborate empirical rules may develop. As the later applications become more varied, more complex rationalist procedures emerge. Boothby and Clough (2017) have previously noted this progression in the design of reinforced concrete. The evolution of the design of a metal roof as a diaphragm also follows this customary path for the adoption of new building technologies. Sheet steel deck began to be applied to flat roofs supported by joists in the 1920s. In one initial application each sheet was stiffened around the perimeter with a small steel angle, which greatly increased the in-plane shear capacity of the sheets – this capacity was accepted empirically, but not calculated (American Builder, 1924). In the further development of the application of sheet metal roofs, the perimeter angles disappeared, but no bracing was added. A 1953 article 'Metal building products…' (American Builder, 1953) shows open web steel joists in use with sheet steel roof deck. The continuing lack of additional bracing shows that the shear capacity of the roof was assumed empirically, although not calculated. Not until the late 1950s and early 1960s, about 30 years after the adoption of this technology, were experiments and calculation methods developed to qualify a roof assembly for the distribution of the calculated lateral forces. Even then, the procedure incorporated many empirical findings uncritically (Luttrell, 1966). The report of Ammar and Nillson (1973) was a synthesis of test results and theory concerning the use of steel panels as a diaphragm. The process was not incorporated into building codes until the SDI published the first edition of its diaphragm design manual (Luttrell, 1981). This manual contains a description of fastener strength and brief descriptions of the determination of the strength and stiffness of diaphragms based on individual fastener strengths, number and spacing of fasteners.

3.7. Solution of routine problems

Empirical design is directly applicable to the routine production of similar designs in building technology. When an element or assembly has been designed multiple times for similar situations, it is no longer necessary to review the theory under which it was designed. An example of this, presented later, is the size of the studs in wood bearing walls. As the size choices are limited and the loads on such bearing walls are consistent from project to project, it may not be necessary to calculate loads and strength explicitly to determine the size of this element.

Designers who work regularly with rectangular bays in commercial steel buildings could choose the beam depth without calculations. The bay size for office buildings is consistently around 8000–9000 mm, the spacing of the infill beams about 2000–2500 mm, and the slab thickness is consistently 200 mm. As the loads are the same for office buildings, the design bending moments vary by less than 20%, so that the depth of the infill beams and the girders are consistent (Box 3.1).

Box 3.1

The design of an uncomplicated bay in a non-composite steel-framed building usually results in a bay, approximately 8 m × 8 m, with infill beams spaced 2 m apart, supporting a 200 mm concrete slab. The following calculations can be performed

Total floor load (including live load) 7500 N/m^2

Floor slab	2500 N/m^2
Steel framing	500 N/m^2
Other dead loads	500 N/m^2
Live load	4000 N/m^2

Infill beams

Total load (1.35D + 1.5L) = 1.35 (3500 N/m^2) + 1.5 (4000 N/m^2) × 2 m = 21.4 kN/m

Bending moment = 21.4 kN/m × (8.0 m)^2 ÷ 8 = 171 kNm

Required Z_x = 171 kNm ÷ 355 N/mm^2 ×1.1[γ_M] = 530 cm^3

Beam size UB 356 × 127 × 33 (Z_x = 543 cm^3)

Span/depth = 22.4

Girders

The girder would be loaded with 171 kN loads at the ¼ points, hence

Bending moment = 684 kNm

Required Z_x = 2120 cm^3

Beam size UB 457 × 191 × 98 (Z_x = 2230 cm^3)

Span/depth = 16.2

The girder design for this system results similarly in a girder proportioned of 16:1 and infill beams at a ratio of 22.4:1 span:depth. The ratios of 24:1 for infill beams and 20:1 for girders has been identified in Chapter 6 of the author's previous work (Boothby, 2018). These ratios can be used securely, at least to choose the nominal depth of the beams in a system.

The repeated design of bolted connections in beams results in an empirical rule that the number of rows of bolts in a simple beam connection can be related to the depth of the beam. The tables in the blue book (SCI, 2022) that give the maximum loading on a beam result in the following maximum shears for various beam depths.

203 mm	73.5–99.4 kN	2
254 mm	80–179 kN	3
305 mm	104–268 kN	4
356 mm	166–318 kN	5
406 mm	223–307 kN	6
457 mm	337–1120 kN	8, up to UB 457 × 152 × 82

On the other hand, a 20 mm diameter bolt in double shear (μ=0.3) is 66 kN, which justifies the approximate number of bolts required by section depth, as shown in the table above.

In non-seismic zones, favourable experience with unreinforced concrete masonry unit (CMU) walls of normal sizes is overlooked in favour of designing to the requirements of ASCE 7-16 (ASCE, 2016) (for determination of loads) and TMS 402 (TMS, 2016) (for assessing strength of masonry) as empirical design was removed from recent editions of TMS 402. Prior to this, the empirical design provisions of the code for structural design of masonry permitted unreinforced 200 mm CMU walls (braced top and bottom) in walls as tall as 3600 mm. The strict application of the wind loading in ASCE 7 and the very small allowable tension stresses normal to the bed joint in TMS 402 place restrictions on the mortar type needed to justify this design. Experience with this wall type indicates they can sustain required maximum wind loads at heights greater than 3600 mm. In this instance, at least, empirical design offered a more effective way to design low-rise unreinforced masonry buildings that compared favourably with experience and unfavourably with the rational provisions of the prevailing codes for wind loading and for masonry design. The basic equivalence of the results of the empirical design by the height:width ratio and the rational design are shown in Box 3.2.

Box 3.2 Wind load calculations of low-rise masonry building by ASCE 7-16 (ASCE, 2016)

Assumptions: 12-feet high (allowable by former empirical design procedure) regular geometry, no terrain or elevation issues, 8-inch unreinforced CMU

Exposure B is most likely for a suburban shopping building

Non-bearing exterior wall considered part of the main wind-force resisting system

Step 1: Risk category II

Step 2: 51 m/s

Step 3: $K_d = 0.85$

Exposure category B

Topographic factor: n/a (1.0)

Ground elevation factor: n/a (1.0)

$G = 0.85$ a rigid building can be assumed for a one-storey masonry building

Enclosure classification = enclosed

$GC_{pi} = \pm 0.18$

Step 4: $K_z = K_h = 0.7$ (h<15 feet)

Step 5: $q_z = q_h = 0.613\ (0.85)\ (0.70)\ (115)^2 = 948\ \text{N/m}^2$

Step 6: Determine external pressure coefficient C_p

From Figure 27.3.1; 0.8 for windward wall

Step 7: Calculate p on windward wall

$p = [0.85(0.8) + (0.18)]948 = 815\ \text{N/m}^2$

The loads from the envelope procedure are not directly comparable as they include the combined effect of pressure on the windward wall and suction on the leeward wall (different walls)

3650 mm high wall subjected to self-weight (1680 N/m^2) and lateral load of 815 N/m^2

Allowable stress design load combination $0.6D + 0.6W$

At mid-height, $M = 0.6\ (1.36\ \text{kNm/m}) = 0.816\ \text{kNm/m}$

$P = 0.6\ (1680\ \text{N/m}^2) \times 1830\ \text{mm} = 3070\ \text{N/m}$

Combined stress at mid-height; face-shell bedded 8-inch unreinforced CMU; $A = 30\ \text{in}^2/\text{ft}$; $S = 81\ \text{in}^3/\text{ft}$

$(0.816\ \text{kN/mm})/(4350\ \text{cm}^3/\text{m}) - (3070\ \text{N/m})/(634\ \text{cm}^2/\text{m})$

$= 0.187\ \text{N/mm}^2 - 0.048\ \text{N/mm}^2 = 0.139\ \text{N/mm}^2$

Compare with allowable stress; type M or S MC mortar 138 N/mm^2

Empirically designed wall slightly exceeds allowable stress if MWFRS (trib. area >65 m^2)

3.8. Empiricist appeals

Empiricist appeals are inevitable in structural engineering design. There are simply aspects of structural design that transcend logic and demand an appeal to experience. The presence of empiricist appeals in contemporary building codes will be described more fully in Chapter 5. The following are examples of such appeals.

The design of hooks on concrete reinforcing bars, such as the 135° hooks on stirrups and ties, has no rational basis – the provisions for minimum bend radius and extension beyond the bend are simply products of experience. The experience includes the manufacturing, fabrication and construction process. The bend minimum radii ensure that the reinforcing bar will not crack or deform excessively during the fabrication process and will further ensure that the hooks are large enough to anchor the bar adequately.

Reinforcement in one-way concrete slabs of a thickness of 150 mm or less is designed to be placed at mid-thickness of the slab or close to this depth in thicker slabs. However, in the construction process: preparation for concreting and concrete placement, this reinforcement is usually trampled to the bottom of the slab. There is sufficient satisfactory experience with one-way concrete slabs with the reinforcement designed to be placed at mid-thickness to make an implicit empiricist appeal that this design procedure works effectively.

In the US, AISC 360 allows the loads applied to the clip angles to be taken as concentric for an eccentricity of 75 mm or less. As the load on the outstanding leg of a clip angle usually has an eccentricity of 40–50 mm, this provision is invoked very frequently. Although this assumption neglects a potentially significant contribution of bending or torque to the stresses in the connections, there is a long history of satisfactory performance of connections designed under this assumption (AISC, 2017: pp. 10–18). The procedures of Eurocode 3 do not include this simplification (SCI, 2014: p. 115).

3.9. Empirical design as a supplement to rational analysis

Empirical thinking can also inform or correct the findings of a rational analysis. The rational analysis of a double top plate for a wood stud bearing wall in residential construction may result in the conclusion that the top plate is insufficient. A simplified version of this calculation is carried out in Box 3.3.

Box 3.3

Load up to 5000 N (concentrated end reaction of a floor joist)

Span up to 600 mm

Midspan bending moment, if load is applied at centre = 750 Nm

Maximum bending stress = 750 Nm ÷ 91.3 cm^3 = 8.2 N/mm^2

The wood species and grades in residential construction (usually spruce-pine-fir no. 2 or equivalent; F_b = 5.3 N/mm^2) rarely have allowable stresses close to 8.2 N/mm^2. The rational conclusion that this top plate may be inadequate is overruled by the empirical fact that this practice is nearly universal.

Empirical design has a significant place in preservation engineering, which will be discussed further in Chapter 7. Not only are historic structures often originally designed using empirical methods, but their design can be difficult to justify by modern methods. It often happens that an old or historic structure does not conform to modern standards of building but has survived effectively over many years. The empirical observation that the structure has endured could be an important part of the assessment of such a structure. This point is explored in detail in Chapter 7.

However, in reviewing historic structures, it is common for engineers to reject empirical evidence of the effectiveness of a structure and rely too heavily on a rational procedure. In the review (Darden and Scott, 2005) of the Wisconsin Avenue viaduct, an 1831 stone arch bridge over the C & O canal, the engineers found that 'Although the bridge was not "overtly" structurally deficient, its load rating could not be determined accurately.' Because of this inability to determine the load rating, the same engineers undertook repairs to the bridge, in spite of the empirical evidence of the bridge's adequacy. All of the load rating and strengthening calculations were done by the manufacturer of the proprietary strengthening system used in the project.

An account is available of strengthening of a historic reinforced concrete arch bridge in the UK (Canning, 2011). The description of the structure includes a similar discussion of the capacity of the bridge

> The original structure had an assessed capacity of 7.5 tonnes assessment live load (ALL) in accordance with BD21. No weight restriction to the bridge was in place as it provides critical access to the village. Regular monitoring of the structure was therefore undertaken until it could be strengthened.

So, the empirical observation is that the bridge functioned acceptably without a weight restriction, but that the analysis of the bridge had concluded that it could not support the loads that it was already supporting.

A better approach to the management of obsolete structural materials is exemplified in Goldyn and Urban (2020). In that study, nineteenth century cast iron columns of uncertain origin, uncertain material properties and variable dimensions were encountered. In spite of these doubtful characteristics, these columns were serving their intended function well. The authors reviewed the design of cast iron columns according to Gordon's formula (a semi-empirical formula used in the 1800s, see Boothby (2015)) and subjected selected columns (from a demolished portion of the building) to mechanical testing, concluding that the columns had sufficient strength for the proposed building re-use and allowed them to remain in place.

The assessment of a structural element that is apparently functioning well is a fundamental question in the assessment of an empirically designed structure. When the structure is not showing signs of distress or is supporting the loads that it is called on to support and the analysis results show that the structure has deficient strength, this can be considered a call to refine the method of analysis to explain why the structure is working, rather than to adhere to a single analysis method that finds that the structure is deficient.

It is noted that the bias against historic structures is waning, as there is a substantial literature on the assessment and repair of historic structures in which measures are taken to ensure that unwarranted repairs do not take place and that, where necessary, repairs respect the historic integrity of the building. On the other hand, the demands of contemporary building codes regarding specificity of design method and general loading are making the assessment of empirically designed structures increasingly difficult.

Beyond these widely distributed ways in which empirical design may be implemented, there are other areas, conscious or unconscious, where empirical design is useful to the engineering profession. Empirical design has further utility in assessing structures with unknown features or in general assessment of a population of structures. The use of simplifications allows the designer to focus on the important parts of the assessment exercise and to leave some of the rationalist calculations for later, when a more detailed investigation may be needed.

Empirical design is especially useful in the preliminary design of buildings. In addition to its usefulness to practising engineers, the ability effectively to determine sizes, materials and configurations for building structures is also very useful in engineering education.

André Coin's book *Ossatures des Bâtiments* (Coin, 1973) describes a set of hybrid rational/empirical rules for the construction of the framing systems for reinforced concrete buildings. It is largely meant as a text for advanced engineering students to describe the applications of reinforced concrete into ordinary building frames. The purpose of this book is similar to that of Boothby (2018), laying down rules that allow preliminary sizes of structural elements in an overall building design.

3.10. Conclusions

Although the design of artefacts in engineering is a fundamentally different practice from the investigations of natural science, both fields use observations of nature and the development of theories to predict natural actions in similar ways. The difference between the two approaches rests with the fundamental purpose of each. Because engineering is constructive, engineers can work with much less background information concerning causes of phenomena and are content to work with theories with considerably less coverage than scientists. As working without a clear theory or ignoring causes are attributes of empiricism, the result is that some sort of empiricism, often in the form of empiricist appeals, pervades the practice of engineering.

Empirical design has significant advantages. These include the production of multiple routine objects from similar materials with similar purposes. When the conditions are very similar and the production rate is high, only the class of solutions requires investigation, while a specific solution can be applied to a range of other solutions. It is also often expedient to have recourse to empirical design in the description and design of objects whose analysis is unreasonably complicated. The complications can be described in proportional terms with reference to the object that is being designed. The behaviour of many widespread time-tested structural systems is complex and designing with reference to the predicted behaviour rather than the widespread practice can result in errors. In such cases, it is customary to refer most solutions to a simplified form of empirical analysis. Other uses of empirical design include the introduction of new materials and new processes, the development of preliminary designs and engineering for historic preservation.

REFERENCES

ACI (American Concrete Institute) (2019) ACI 318-19: Building code requirements for structural concrete. ACI, Farmington Hills, MI, USA.

AISC (American Institute of Steel Construction) (2016) AISC 360-16: Specification for structural steel for buildings. AISC, Chicago, IL, USA.

AISC (2017) *Steel Construction Manual*, 15th edn. AISC, Chicago, IL, USA.

American Builder (1924) What's New? Steel for the flat roof. *American Builder* **37(4)**: 132.

American Builder (1953) Metal building products: Steel joist design tables. *American Builder* **75(4)**: 329.

Ammar AR and Nillson AH (1971) *Analysis of light gage steel shear diaphragms*. Center for Cold-Formed Steel Structures Library, 84. https://scholarsmine.mst.edu/ccfss-library/77/ (accessed 24/05/2023).

ASCE (American Society of Civil Engineers) (2016) ASCE 7-16: Minimum design loads and associated criteria for buildings and other structures. ASCE, Reston, VA, USA.

AWC (American Wood Council) (2018) *National Design Specification*. AWC, Leesburg, VA, USA.

Bonowitz D and Maison D (2003) Northridge welded steel moment-frame damage and its use for rapid loss estimation. *Earthquake Spectra* **19(2)**: 335–364.

Boothby TE (2015) *Engineering Iron and Stone: Understanding Structural Analysis and Design Methods of the Late 19th Century*. ASCE Press, Reston, VA, USA.

Boothby TE (2018) *Empirical Structural Design for Architects, Engineers and Builders*. ICE Publishing, London, UK.

Boothby TE (2020) Empirical design of masonry arch bridges. *Journal of Architectural Engineering* **26(1)**: 4.

Boothby T and Clough S (2017) Empiricist and rationalist approaches to the design of concrete structures. *APT Bulletin* **48(1)**: 6–14.

BSI (2004) BS EN 1992-1-1-2004: Eurocode 2: Design of concrete structures – Part 1–1: General rules and rules for buildings. BSI, London, UK.

BSI (2005) BS EN 1993-1-1:2005: Eurocode 3: Design of steel structures – Part 1–1: General rules and rules for buildings. BSI, London, UK.

Butler (2023) *MR-24 Roof System*. 2023. https://www.butlermfg.com/products/roof-systems/mr-24/ (accessed 19/03/2023).

Canning L (2011) Minsterley Bridge strengthening using novel methods. *Proceedings of the 5th International Conference of Advanced Composites in Construction*, ACIC 2011, pp. 21–29.

Cartwright N (1983) *How the Laws of Physics Lie*. Oxford University Press, Oxford, UK.

Coin A (1973) *Ossatures des Bâtiments*. Edition Eyrolles, Paris, France.

Darden C and Scott TJ (2005) Strengthening from within. *Public Roads* **68(5)**: 8.

Goldyn M and Urban T (2020) Failures of the cast-iron columns of historic buildings – case studies. *Infrastructures* **5(71)**: 18.

Hayes L (1915) *Empirical Design*. Carpenter & Co., Ithaca, NY, USA.

Jorquera N, Vargas J, de la Luz Lobos Martínez M and Cortez D (2017) Revealing earthquake-resistant geometrical features in heritage masonry architecture in Santiago, Chile. *International Journal of Architectural Heritage* **11(4)**: 519–538.

Koen BV (2013) Debunking contemporary myths concerning engineering. In *Philosophy and Engineering: Reflections on Practice, Principle and Process* (Michelfelder *et al.* (eds)) Philosophy of Engineering and Technology 15, Springer, Dordrecht, The Netherlands.

Lee K and Foutch D (2002) Performance evaluation of new steel frame buildings for seismic loads. *Earthquake Engineering and Structural Dynamics* **31(3)**: 653–670.

Luttrell L (1966) *Structural performance of light-gage steel diaphragms*. PhD dissertation, Cornell University, Ithaca, NY, USA.

Luttrell L (1981) *Diaphragm Design Manual*. Steel Deck Institute, Fox River Grove, IL, USA.

Miller DK (1998) Lessons learned from the Northridge earthquake. *Engineering Structures* **20(4)**: 249–260.

Mora Alonso-Muñoyerro S, Rueda Marquez de la Plata A and Cruz Franco P (2014) Consolidation of historical masonry: Past experiences and future forecast. *Construction and Building Research* 299–303.

Rankine WJM (1865) *A Manual of Civil Engineering*, 4th edn. C. Griffin, London, UK.

Romel G, Sherif L and Ashour S (2020) The history of monasteries in Egypt as self-sustained. In: *13th International Conference on Civil and Architecture Engineering (ICCAE-13)*, 7–9 April 2020, Cairo, Egypt (11 pp.); ISSN: 1757-8981; DOI: 10.1088/1757-899X/974/1/012017

SCI (Steel Construction Institute) (2014) *Joints in Steel Construction: Simple Joints to Eurocode 3*. SCI, Ascot, Berks., UK.

SCI (2022) *The Blue Book*. https://www.steelconstruction.info/images/b/b7/SCI_P363.pdf (accessed 23/07/2022).

SDI (Steel Deck Institute) (2004) *Diaphragm Design Manual*. SDI, Fox River Grove, IL, USA.

Staples M (2014) Critical rationalism and engineering (ontology). *Synthèse* **191**: 2255–2279.

Staples M (2015) Critical rationalism and engineering (methodology). *Synthèse* **192(1)**: 337–362.

Structural Engineering Solutions (2022). https://learnwithseu.com/flexible-vs-semi-rigid-vs-rigid-diaphragms/ (accessed 23/07/2022).

TMS (The Masonry Society) (2016) TMS 402: Building code requirements and specifications for masonry structures. TMS, Boulder, CO, USA.

Trautwine JC (1874) *The Civil Engineer's Pocket-book*. Claxton, Remsen, and Haffelfinger, Philadelphia, PA, USA.

Boothby T
ISBN 978-0-7277-6633-5
https://doi.org/10.1680/edse.66335.033

Chapter 4
Purely empirical builders and their products

The history of construction furnishes examples of the success of empirical builders. Although, even from ancient times, natural philosophers contributed to the rational theory of mechanics, their ideas are not applied to the design of buildings until the seventeenth century and, at that time, only in very exceptional circumstances. Effectively, the design of buildings remained empirical into the nineteenth century. We do see applications of the study of mechanics to building in the nineteenth century. However, the prevailing mode in the early nineteenth century is empirical, while in the second half of the nineteenth century, there is a balance between empirical and rational design.

4.1. Empirical design in antiquity

The architecture of ancient Greece and Rome is an early example of the success of empirical design. In addition to the written evidence of the use of proportions among authors such as Vitruvius and Frontinus (Figure 4.1), structures exhibiting similar proportions are widespread throughout the Roman empire and are distributed over several centuries. Examples of such use of proportioning are the proportions of Greek and Roman temples and the proportions of Roman bridges.

The configuration and design of structures for buildings in the ancient world appears to be a combination of the experience of individual builders and the proportional rules that derive from the combined knowledge of all builders. Our knowledge of these building practices comes from a limited number of sources. Foremost among these sources are the buildings themselves. Although it is not possible to infer exactly what proportional rules these builders may have followed, the adherence of buildings of various types – residences, basilicas, bridges, temples, towers – to proportional and constructive rules is apparent. Moreover, these practices often appear to be stable through time. Figure 4.2 shows three Roman multi-span bridges from different centuries, located in different countries. The spans of each of the arches in the bridges vary between approximately 7 and 9 m. In spite of the differences in location and time, the bridges have very similar features: multiple arches slightly less than a semicircle so, with the same span:rise ratio, the proportion of ring thickness to inner radius of the arch, the proportion of arch opening to pier width, the slight stilting of the arches on the piers, and the opening present in the spandrels in each of the interior piers. These features indicate proved construction practices for different span lengths, differing numbers of spans and different roadway heights. Similar conclusions would result from a comparative study of the proportions of Greek and Roman temples or of Roman baths and basilicas.

Figure 4.1 A plate from Vignola (17th century) describing the proportions of the classical orders of architecture. Although buildings from antiquity do not strictly follow the proportions described in later centuries, the use of proportions for the design of buildings in antiquity is well known

Moreover, a small number of documents describe the construction of buildings in the ancient world. Although Frontinus (1993) was concerned with the construction of aqueducts below and above the ground, he gives no details of their construction, speaking more broadly on sources of water, distribution of water, detection of fraud and other matters. His discussions of the aqueducts existing at the time include a simple statement of what proportion of each of the major aqueducts is above the ground on 'opus arcuatum' (arched work). The implication is that the construction of arches for aqueducts is routine, following accepted practices and

Figure 4.2 A trio of Roman bridges. Although these bridges were built in different centuries and in different Roman provinces, they all observe similar proportions and similar construction features. (a) Pont Tibère, Sommières, France, first century BCE to first century CE (Created by Carole Raddato under a Creative Commons Attribution and ShareAlike licence, https://creativecommons.org/licenses/by-sa/4.0/deed.en); (b) Ponte Romano Mérida, Spain, late first to early second century CE (Created by Irene Piergentili under a Creative Commons Attribution and ShareAlike licence, https://creativecommons.org/licenses/by-sa/4.0/deed.en); (c) Ponte de Vila Formosa, Portugal, second to third century CE (Created by Fernando LB Maria under a Creative Commons Attribution and ShareAlike licence, https://creativecommons.org/licenses/by-sa/4.0/deed.en)

(a)

(b)

(c)

experience, a point which is borne out by examination of Roman above-ground aqueducts of ordinary types. Frontinus also draws on his large experience with water distribution to recognise the proportional rules between the quantity of water distributed and the size of the pipe that effects this delivery. His use of units of a digit or an ounce as measures of the diameter of pipes, or squared units as a measure of the area of a pipe, all reflect the importance of proportional rules. In this case, the area of a pipe or adjutage cross-section is in proportion to the volume of water delivered.

The first century (CE) writings of Vitruvius (1998) apply to the structural design of buildings, and nearly all the cases where such matters are discussed describe proportional rules. Vitruvius' descriptions of the height:diameter ratios of columns, ranging from 6:1 (Doric) to 10:1 (Corinthian) are well known. Such rules can be recognised as a means of assuring the stability of an important feature of a stone structure. Like many of Vitruvius' rules, the proportioning rules for columns could equally be considered human proportions, as Vitruvius represents them, a requirement for structural stability, or a judgement of visual appeal. These three qualities are inseparable in Vitruvius' conception of design. The intercolumniations are presented in the form of a proportion, ranging from 1.5:1 (pycnostyle) to 3:1 (diastyle) and the implications of this ratio for the choice of material for the architrave is given similar weight.

Not only in building, but also in the construction of machines does Vitruvius rely on proportional ratios. In the preface to the entire work, he draws an analogy between parts of a building, which correspond proportionally to a given module, and parts of a ship, which depend on the distance between oars and the parts of a ballista, in which the entire object is built with reference to the aperture. In the discussions of the construction of a war machine, the size of the aperture is the module that determines the size of each of the parts of the machine. The size of the remainder of the parts, described in great detail, follows from the initial module, which is itself based on the size of the projectile. An example of this treatment follows.

> When a ballista is to shoot a stone weighting two pounds, the aperture in the frame will be 5 digits, with four pounds 6 digits....... When the size of the hole is determined (as the module), let the cross-piece, scutula, which in Greek is called peritretos, be drawn; its length 8 holes, the breadth is to be 2 1/6 holes. The cross-piece when drawn is to be divided along the middle of the line and when the middle is divided, the ends of the figure are to be contracted, so that it is bent obliquely to the extent of 1/6 the length, and 1/4 the breadth where the rope turns.

On the other hand, there is very little explicit science or discussion of causes in the Roman texts, or in the Greek texts that preceded them. There are occasional references to weights or forces or stability, such as the general discussion of machines in Book X, Chapter III of Vitruvius. Some Greek authors hint at structural effects while discussing other matters; for example, Aristotle (1996) (*Physics* Book VII, Chapter IV) states that the cause of motion can be the lack of obstacles as when one removes a column, while Empiricus (1936) (*Against the Physicists* I.229) states, 'Again, the pillar is motionless and the lintel also is motionless. But one should not say that the lintel is motionless because of the pillar any more than the pillar because of the lintel; for when the one is removed the other tumbles down.' In general, though, ancient texts have little recourse to arguments about how buildings fail or to the determination of principles to prevent such effects, or general constructive arguments about how to design or build structures.

It is also possible to infer effective proportional rules used in the design of buildings in late Roman and early medieval times. Some of these rules look back to Vitruvius, while others are original. Vitruvius' discussion of intercolumniation ratios (the ratio of the space between columns to the column diameter) indicates that a ratio of approximately 2 to 2.5 is the maximum that can be supported by an architrave, a solid stone placed directly on the top of the columns. In the early Christian period, we find validation of Vitruvius' ideas. An examination of the interior colonnades in Roman churches from about 500 CE to about 1100 CE shows three categories of intercolumniation ratios (Rodriguez Asilis, 2016)

- columns supporting architrave, associated with intercolumniation ratio of 2:2.5
- columns supporting relieving arches over an architrave, associated with intercolumniation ratio of 2.5:3.0
- columns supporting arcades, associated with intercolumniation ratio of 3 and greater.

A proportional rule can be inferred by examining the bell towers on early Christian and medieval churches in Rome. The proportion of height to width for these structures ranges from 4:1 to 7:1, while most have a height–width ratio of close to 6:1. This proportion can be regarded in the context of Vitruvian symmetry as the proportions of a Doric column, or the height of a man compared with the length of his foot.

Church architecture in the period from approximately 1000 CE to 1500 CE, built according to the Romanesque and Gothic systems of construction, are a particularly fertile ground for the study of the application of proportioning systems in construction. During these time periods, the use of proportioning is applied more through geometry than through arithmetic, as very complex procedures were developed for working with the geometric complexities of producing a vaulted stone structure. The Romanesque style also made significant use of proportions as a method of design, both in plan and section and are reflected in buildings that continue to function as churches. In the Lombard Romanesque system, the plans from church to church are consistent – with two aisle chapels corresponding to each nave bay and a double system of large and small piers. Proportional rules about the shape of the buttresses can also be inferred (Rodriguez Asilis *et al.*, 2018). In Romanesque architecture in France, from approximately 1000 to 1300, there were regional 'schools' that each evolved a separate empirical construction typology. These construction types were characterised by overall nave designs that contributed to the resistance to the vault thrusts (Frankl, 1918). Examples are the Potevin school that used elevated aisles (Figure 4.3(a)), forming a so-called hall church, the Auvergne school that used galleries in the aisles as buttresses, the Aquitaine school that built mutually buttressing cupolas (Figure 4.3(b)) and the Norman school that buttressed barrel vaults, banded or not, with quarter-circle aisle vaults.

Gothic architecture, from the late Middle Ages (1140 to *c.* 1500) displays a complex and detailed approach to the application of proportions to the design of structures. The application of these proportioning (and geometric) rules manifestly resulted in successful buildings. In this case, proportioning is used in the plan, the section, for the complex details of window tracery,

Figure 4.3 Examples of Romanesque schools in France: (a) Notre Dame la Grande, Poitiers, twelfth century. Representative of the Potevin hall churches (Created by Gerd Eichmann under a Creative Commons Attribution and ShareAlike licence, https://creativecommons.org/licenses/by-sa/4.0/deed.en); (b) Cathedral of St Pierre, Angoulême, twelfth century. Representative of the domed churches of the Aquitaine school (Created by Gautre under a Creative Commons Attribution and ShareAlike licence, https://creativecommons.org/licenses/by-sa/4.0/deed.en)

(a)

(b)

interior elevations, the configuration of ambulatories and other details. Most of the practice of proportioning in this period is accomplished by detailed geometric constructions.

Examples of Gothic proportioning include discussions of the general proportioning of a church. The archives of the construction of the Cathedral of Milan are very often cited in this context, as the documents maintained from the start of construction in 1386 to the early fifteenth century contain repeated discussions of the correct proportions for the church. It has been noted that the evolution of these proportions as the project went on tended to lower the overall height of the church, in accordance with structural necessity. When choosing a height based on the triangle or based on the square, the lower relative height of the triangle was chosen. This triangular scheme in section was subjected to modifications that resulted in the height of the crossing being even lower.

The fourteenth century cathedral in Florence, on the other hand, is an example of a church built according to a square. The overall width of the nave and the two aisles and the overall height of the top of the dome are approximately equal at 144 Florentine *braccia*, or 83 m (Trachtenberg, 2001). Harmonic ratios are respected throughout this building, in plan, where the four large nave bays correspond to the aisle width of one-half of the nave bays. The height of the nave is approximately equal to the width of the nave with its two aisles. The nave bays are all square, as are the transept bays.

More detailed and often more explicit rules were applied to the overall proportioning of the cross-section of late German Gothic buildings. Such proportions, applied by experienced builders, where known to provide effective buildings. The nature of these rules, which were recorded in documents, is described by Coenen (1989) (see Figure 4.4). The author proposes a series of plans and sections for late Gothic cathedrals based on the 'Unterweisungen' written in 1516 by Lorenz Lechler. The work by Lechler offers proportional rules for churches of various descriptions and various sizes, including single-naved churches, churches with a nave and two low aisles and 'Hallenkirche', with the height of the nave and the aisles equal. He also describes the proportions of piers, wall buttresses and flying buttresses along with the conditions under which flying buttresses are necessary.

Geometry was used as a means of laying out smaller elements of medieval structures as well. Although the geometrical constructions may not be using proportions explicitly, such geometrical constructions result in proportional ratios between the elements of a construction. The clearest example of this is the late medieval discussion of the layout of a pinnacle by Matthias Roriczer (1965), in which, by geometrical construction beginning with the proportions of the base for a pinnacle, the author succeeds in configuring an entire pinnacle on the basis of simple geometrical constructions.

Similar examples have been inferred from the layout of features of medieval architecture, based on inferences of compass and division-based constructions of elements such as tracery in windows, the section of columns, vault ribs and other features. Such constructions are fully described by Bork (2011).

The success of the medieval way of thinking about structures is exemplified by the fifteenth to seventeenth century 'crown steeple' at St Giles' Cathedral in Edinburgh. A photograph of this feature is shown in Figure 4.5. The open form of the ribs of this steeple reflects the experience with vault forms conducted by medieval architects over several preceding centuries. Curiously, as shown in Figure 4.6, this masterwork of empirical design overlooks a statue of David Hume, the empiricist philosopher.

Figure 4.4 Excerpt from Coenen (1989), using a Master's book of relative dimensions to construct a church ground plan. Examples: nave width = choir width; aisle and ambulatory width = 2/3 nave width; nave bay length = 2/3 nave width; wall thickness = 1/10 nave width; nave pier – octagonal with each face 1/10 nave width (overall dimension $(1 + \sqrt{2})/10$ nave width)

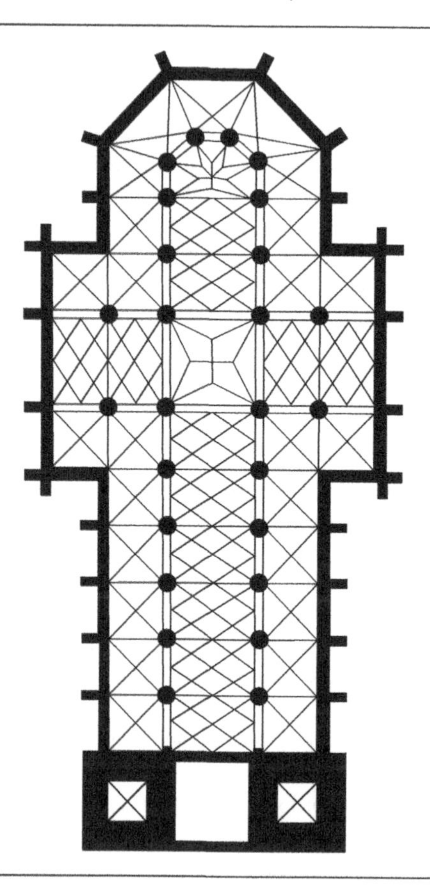

4.2. Renaissance

In the later Renaissance, there is the emergence of writings on mechanics that have some application to buildings. In spite of this, the design of buildings remains largely empirical and largely based on the selection of appropriate proportions. As a variety of authors left treatises on architecture, it is possible to review this evidence and to infer the use of both proportions as an empirical design tool and the recourse to experience as a design and construction method.

The architects of the Renaissance were simultaneously rediscovering ancient Roman architecture and inventing their own. As a result of this, the empirical understanding of the Italian Renaissance architects is a combination of ancient Roman experience combined with new discoveries. The writers of architectural treatises, such as Alberti and Palladio, pay homage to Vitruvius in one way or another (Alberti, for instance, in entitling his treatise *Ten Books on Architecture* (Alberti, 1988)). Their descriptions of materials and their uses parallel Book II in Vitruvius, and the descriptions of the classical orders also depend on Vitruvius. Palladio's

Figure 4.5 Crown steeple, St Giles' Cathedral, fifteenth to seventeenth century, Edinburgh, UK

Figure 4.6 David Hume, the empiricist philosopher with a masterpiece of empirical design in the background (Photograph by author)

description of walling is taken directly from his observation of Roman construction (Palladio, 1997). On the other hand, both Alberti and Palladio insert their own ideas and up-to-date techniques into their descriptions of the art of building.

The later part of the Renaissance, the late seventeenth to the eighteenth centuries, saw the emergence of rational theories of mechanics and their occasional application to building technology. Among the most well known among these theories are Robert Hooke's discussion of the mechanics of the arch and Poleni's application of these theories to the analysis of the dome of St Peter's basilica in Rome. These are discussed in many sources, including Heyman (1999) and Benvenuto (1991). Similarly, significant progress was made in the analysis of beams, leading to the development of a theory of flexure that is still used today.

However, structural design remained generally empirical in application. Beams were primarily designed using empirical rules, and arches and other masonry structures were designed based on correct proportions. There are significant treatises on construction from the Renaissance, including those of Alberti, Palladio and others, which, in describing the design of beams or arches, rely almost entirely on empirical ratios. Baldi (2011) has an extended description of arches, trusses and beams, and the mechanisms of failure and how to avoid failure for these types of members. The author does not offer any numerical calculation of strength, but relies on ratios – a beam twice as thick may be twice as strong – or ratios for the height: span of an arch.

The major work on architecture of Alberti (1988), *De re aedificatoria libri decem*, includes frequent appeals to experience, and these appeals often take the form of proportional rules. Two examples of his approach to proportioning are presented here: bridge arches and columns. Arches and arch bridges are described in two different parts of the work, arches in Book I, Chapter XII

> But in all openings over which we mark arches, we should contrive to have the arch never less than a half circle, with an addition of the seventh part of half its diameter [span:rise 2, span:thickness 14].

The construction of arch bridges appears in Book IV, Chapter VI (this description is a hybrid of appeals to experience and proportioning)

> But if by the disposition of the piers, the semi-circle should rise so high as to be inconvenient, we may make use of the scheme [segmental] arch, only taking care to make the last piers on the shore the stronger and thicker… and there should not be a single stone in the arch but what is in thickness at least one tenth part of the chord of the arch; nor should the chord itself be longer than six times the thickness of the pier, nor shorter than four times.

Columns are described in Book VI, Chapter XIII thus

> The diameter you divide into four-and-twenty parts, one of which you give to the height of the fillet, which height we mark upon the wall with a small stroke; then take three more

> of those parts, and at that height make a mark in the axis of the shaft, which is to be the center of the next diminution and through this center draw a line exactly parallel with the diameter of the bottom of the shaft, which line must be the diameter of the lower diminution, and be one seventh part shorter than the diameter of the bottom of the shaft....

Palladio's *Four Books on Architecture* (1997) refers to Vitruvius and to measurements of classical architecture, along with advice to contemporary builders on the correct way to build. This advice consists of a combination of rules derived from experience and proportional rules. For instance, in the first book, Chapter XI, 'On the Diminution of Walls'

> However much walls increase in height, so much must they be diminished in thickness: those walls that begin at ground level will be narrower than the foundation by one-half, and the wall for the next floor will be narrower than this by the first half of a quarter, and so on [one additional eighth part of the bottom wall] to the top of the works: carried out with discretion so that the upper walls are not too thin.

In Book III, Palladio offers designs of a truss speaking in detail of the construction of the truss, including the longitudinal beams, the posts, or 'colonnelli', the inclined top chord and web members that standing opposed to each other hold up the whole. Palladio appears to be using an overall depth of approximately 1/8 of the 100 ft span (a modern proportion for a truss). The dimensions of the principal members of the truss are given as a depth of less than a foot and a width of three-quarters, which are similarly suitable proportions, compared with a modern timber truss.

4.3. Nineteenth century

In the nineteenth century, rational design emerged as a useful method for the design of buildings. The number of writings on the topic multiplied and the training of engineers in polytechnic schools began to promote the application of theories of mechanics and the use of reasoning as a tool for the configuration and sizing of elements in buildings. Nevertheless, a significant component of empirical design remained as an important tool for structural design. By the end of the century, structural designers balanced the application of experience and the use of theories of mechanics for the proportioning of elements in a building structure.

The nineteenth century furnishes many examples of the adoption of new materials and new techniques. We have previously described this model for the adoption of new technologies in the context of steel roof purlins and in the use of diaphragm action for galvanised steel roof decks. We have noted that most of these new ideas are often applied in advance of any rational explanation of how the technology works, or of how prototypes are to be designed rationally – that is, the initial use of a new building technology is often empirical, and there is a time lag in the development of rational procedures for the application of this technology, as the profession notes this gap in application and tries to fill it. The initial rational models are also in themselves sketches of a complete rational method, in large part empirical, especially as the new application is complicated and needs significant effort to provide a complete rational method to justify the analysis of the material.

Examples of the introduction of new materials include the application of reinforced concrete (Boothby and Clough, 2017). The initial experiments with this material took place in the mid-nineteenth century. Within two decades of the first adoption of this material, competing systems of concrete construction presented their own semi-rational method of design. Two or three decades elapsed before any sort of universal understanding of the mechanics of reinforced concrete was achieved, and additional time beyond this to develop procedures for structural design for this material. A further example is the use of wood trusses. Wood trusses are in evidence in construction since Roman times and were particularly widely used in the Renaissance. Both Palladio (1997) and Baldi (2011) offer explanations of the construction or the mechanics of a truss; however, neither description is of any value in sizing elements: Baldi describes how interference between elements locks the truss against significant deformation. The empirical rules available for this structural type relate primarily to the configuration of the truss, without offering advice on what size to make the elements of the truss. It is only in the early mid-nineteenth century that a widespread method of estimating the forces in the elements of a truss becomes available, and effectively, not until the end of the century that the analysis of a truss becomes regularised into a design procedure.

The author (Boothby, 2015) includes descriptions of the practice of empirical forms of analysis for all of the major structural materials in the late nineteenth century: wood, masonry, iron or steel, and for configurations such as girders, beams, joists and trusses. This work shows a significant reliance on empirical rules and experience-based design in all aspects of construction, even as rational methods of analysis were increasingly widely available.

Load-bearing brick masonry, for instance, was designed by a widely applied rule of starting from a 12 in minimum wall thickness and of adding one wythe of brick for each two floors supported. Thus, a four-storey building would have a 16 in wall at the base and a 12 in wall for the top two storeys. The application of this rule was carried to its limit in the 16-storey Monadnock Building in Chicago, with bearing walls that are approximately 5 ft thick at the ground floor.

Empirical design was continuously practised in the nineteenth and into the twentieth century for the design of wood load-bearing structures. Available in Boothby (2015) are published empirical rules and procedures for the design of wood posts and wood beams. There is also a significant incursion of rational design into the analysis of wood structures in the late nineteenth century; however, the rational design procedures for finding stresses in floor joists, for instance, are reduced to the form of empirical rules: rules that do not require invoking the actual stress analysis and that may include mixed or inconsistent units, approximations, or other empirical features.

Among the basic structural materials of masonry, wood and iron or steel, the design of iron and steel in the late nineteenth century tended the most to follow rational analytical procedures. Some of them would be recognisable to a contemporary engineer, while other of these procedures take a special effort to follow. As with wood structures, there is a tendency to reduce the results of the analysis of a steel beam, for instance, to empirical rules. In particular, the direct calculation of the required area of the flanges of steel lintel beams proposed by Kidder (1886) is an example of the reduction of an analytical procedure to an empirical rule (Boothby, 2015: p. 53).

Columns were designed to the early twentieth century by 'Gordon's formula', which is a semi-empirical analytical method for the maximum compressive stress supported by a column. The formula depends on empirical parameters, such as the application of different factors for different cross-sectional shapes, but the formula is robust in its estimation of buckling loads on metal columns.

4.4. Conclusions

An important observation emerges from this very brief survey of construction practices over approximately 20 centuries. Many of the structures from the past that we admire greatly were designed entirely, or almost entirely, on the basis of experience, either direct personal experience of the builder with similar structures or experience as handed down verbally, or in the form of empirical rules. It is difficult to overlook the success of this method of designing structures: successful experience with a construction type engenders further successes and further experiments and, finally, an important and enduring construction type is established. It may also happen, as it did during the Renaissance, that the builders of a later age try to recover the experience of a previous age. Renaissance architects, such as Alberti, Serlio, Palladio and others visited (or excavated) Roman ruins to infer the rules of building that guided the design of these structures. The treatises on architecture that these investigators wrote, along with their re-reading of Vitruvius and other Roman works, discuss the rules used in Roman construction, and use similar rules to apply to the design and construction of buildings in their own age. A similar effort may be seen in the early nineteenth century Gothic revival, in which books of measured drawings of Gothic buildings were widely circulated and used as models for contemporary construction.

Such wisdom is less common in the present age. In many cases, well-established rules of construction have been replaced by rules of analysis. In contemporary engineering the general principles of structural analysis may be applied to a masonry arch, for instance, but they do not necessarily provide the productive information that is furnished by experience, and the simplifications necessary to analyse a structure often make the structure appear to be weaker than it is. This can give rise to situations such as those outlined at the beginning of Chapter 3 in which the analysis of an arch bridge that has no apparent defects can result in a decision to strengthen the bridge. The builders of previous ages, as discussed in the present chapter, were usually more knowledgeable about the construction rules of prior ages. Alberti's treatise draws on the successes of Roman architecture and engineering and was used to guide engineers, and Pugin draws on the success of the Gothic architects and his work is used for the guidance of architects. This is an example that contemporary engineers may do well to follow.

REFERENCES

Alberti L (1988) *On the Art of Building in Ten Books*. tr. Rykwert J, Leach N and Tavernor R. MIT Press, Cambridge, MA, USA.

Aristotle (1996) *The Physics*. tr. Wickstead P and Cornford F. Harvard University Press, Cambridge, MA, USA.

Baldi B (2011) *In Mechanica Aristotelis Problemata Exercitationes*. Berlin: Edition Open Access. https://edition-open-sources.org/sources/4/index.html (accessed 26/05/2023).

Benvenuto E (1991) *An Introduction to the History of Structural Mechanics. Part II. Vaulted Structures and Elastic Systems*. Springer, New York, NY, USA.

Boothby T (2015) *Engineering Iron and Stone: Understanding Structural Analysis and Design Methods of the Late 19th Century*. ASCE Press, Reston, VA, USA.

Boothby T and Clough S (2017) Empiricist and rationalist approaches to the design of concrete structures. *APT Bulletin* **48(1)**: 6–14.

Bork R (2011) *The Geometry of Creation*. Ashgate, Surrey, UK.

Coenen U (1989) *Die spätgotischen Werkmeisterbücher in Deutschland als Beitrag zur mittelalterlichen Architekturtheorie*. Verlag Günter, Mainz, Germany. (In German.)

Empiricus S (1936) *Against Physicists*. tr. Bury R. Harvard University Press, Cambridge, MA, USA.

Frankl P (1918) *Die Baukunst des Mittelalters*. Akademische Verlagsgesellschaft Athenaion, Berlin, Germany. (In German.)

Frontinus SI (1993) *The Stratagems and the Aqueducts of Rome*. tr. Bennett C. Harvard University Press, Cambridge, MA, USA.

Heyman J (1999) *The Science of Structural Engineering*. Imperial College Press, London, UK.

Kidder FE (1886) *The Architects' and Builders' Pocket-book*, 3rd edn. Wiley, New York, NY, USA.

Palladio A (1997) *Four Books on Architecture*. tr. Tavernor R and Schofield R. MIT Press, Cambridge, MA, USA.

Rodriguez Asilis Y (2016) *Use of proportions as a structural design tool in early Christian and early medieval churches*. MS thesis. Pennsylvania State University, PA, USA.

Rodriguez Asilis Y, Cardani G, Coronelli D and Boothby T (2018) Proportional design of Lombard buttresses. *10th International Masonry Conference, Milan, Italy*.

Roriczer M (1965) *Das Büchlein von der Fialen Gerechtigkeit: Faksimile der Originalausgabe Regensburg 1486*. Pressler, Wiesbaden, Germany. (In German.)

Trachtenberg M (2001) Architecture and music reunited: a new reading of Dufay's Nuper Rosarum Flores and the cathedral of Florence. *Renaissance Quarterly* **54(3)**: 740–775.

Vitruvius MP (1998) *Vitruvius on Architecture*. tr. Granger F. Harvard University Press, Cambridge, MA, USA.

Boothby T
ISBN 978-0-7277-6633-5
https://doi.org/10.1680/edse.66335.047

Chapter 5
The conventions and codes of structural engineering

5.1. Introduction

In this chapter, we first investigate the conversion of structural codes from an allowable stress design basis to a load and resistance factor design (in the US) or limit states (in Europe) basis. We identify the idea of conventionalism that underlies the adoption of the new codes. We then describe other aspects of code-based design that are primarily conventional or reflect empirical understanding. The empiricist appeals that are present in many structural design decisions receive attention near the end of the chapter.

Conventionalism, according to Runes (1983) is

> any doctrine according to which *a priori* truth, or the truth of propositions of logic, or the truth of propositions (or of sentences) demonstrable by purely logical means, is a matter of linguistic or postulational convention (and thus not absolute in character).

The mid-late twentieth century furnishes an important example of conventionalism in structural engineering, reflected by an overall change in the design of structural systems. The conversion from an allowable stress design basis to a strength design, load and resistance factor design, or limit states design basis has unfolded over decades, led by concrete and steel material codes. The conventionalist ideas in building codes become evident during conversion to new codes, as a different set of conventions are adopted for the analysis and design of a structure, without really altering the overall conception of structural behaviour.

The evidence of the codes themselves supports the idea that the treatment that structural engineering is based less on a rationalist approach to the determination of loading or the sizing and configuration of elements of a structure. Instead, the codes are governed by the ideas of conventionalism and empiricism, both of which are reflections of a non-rationalist view of the behaviour of a structure. In the conventionalist viewpoint, the description of structural actions follows agreed-upon conventions and the explanations of phenomena that incorporate these actions are also made to conform to these conventions. In an empiricist viewpoint, as has been previously described in Chapters 1 to 3, the requirements for a structure are not determined by reasoning through the scientific understanding of the behaviour of a structure, but by simple prescription of the rules for the design of the structure or the element of the structure in question. The deference to code-based, conventional methods of analysis reflects the understanding presented in Chapter 3 concerning the philosophy of engineering. An engineer's concern with creating artefacts allows them to bypass understanding underlying causes or to bypass the use of more comprehensive theories.

The basis of building codes, as they have evolved through the development of the profession, is conventionalist. Evidence for this assertion is the willingness of the writers of the codes to discard one set of conventions of the design of structures for another convention. This is done without changing the basic premise of the analysis and design of a structure, only the prescription of the rules used to accomplish this design. The primary example that will be discussed is the conversion of most of the design codes in the US and in the UK from an allowable stress design basis to a 'load and resistance factor design' (LRFD) basis in the US or a 'limit state design' (LSD) basis in the UK. Other modifications to these codes, such as the resetting of the load factor for wind and earthquake loads to 1.0 in the US, speak further to the idea that the building codes represent agreed-upon conventions rather than any more fundamental rationalist procedure for designing structures.

The procedures used to analyse structures are also the product of convention, from the rules of material behaviour to the methods used for structural analysis. Of particular note are the adoption of elastic methods of analysis, even when the theory of failure (concrete and steel) is elastic.

Two types of building codes will be described here. The first type of code is a general code prescribing the loadings: live load, wind load, flood load, earthquake load, that need to be used in structural design. We will refer to these as loading codes. The examples are ASCE 7: Minimum design loads for buildings and other structures (ASCE, 2016) and Eurocode 1: Actions on structures (BSI, 2002b). The second type of code is specific to structural materials: wood, masonry, steel or concrete, as, for example ACI 318: Building code requirements for structural concrete (ACI, 2019) or its counterpart Eurocode 2: Design of concrete structures (BSI, 2004a).

Throughout, these codes appear to be based on rational analysis. The loading codes, such as ASCE 7 in the US or Eurocode 1 in the UK, provide methods of assessing wind loads on structures departing from the stagnation pressure derived from Bernoulli's equation for fluid flow, by a rationalist procedure. The discussion of wind loads then proceeds through distribution of wind loads based on experiments on distributions of pressures and suction to walls and roofs. The methods of the code permit a rational calculation of the design wind pressures on every part of a structure. Similarly, for earthquake design, the development of loading begins with a determination of the fundamental period of vibration of the building, and the application of a site-specific simplified response spectrum for the building to determine the major loads on the building frame.

Design floor loads, on the other hand, are adopted by convention. There is little evidence that the floor loads used in modern practice differ from those used throughout the late nineteenth and early twentieth centuries, nor any persuasive evidence that these loads result from rational analysis of the loading environment.

In the material codes, the basic premises of the analysis, say, of a reinforced concrete section are provided. Methods of analysis to determine shear and bending moment on a structural element are described. A basis for the strength of the section is determined. These procedures are based on notionally rational procedures developed, for reinforced concrete, over the past century.

A more detailed examination of these codes that follows shows that, in addition to rational analysis, they contain a significant accumulation of conventional procedures, empiricist procedures, empiricist analysis and empiricist appeals. Many of these features are unavoidable in design and most of them are the result of decades of experience. In the following, we describe significant conventionalist and empiricist features of the codes that are used by the engineering profession. This discussion is not intended as a critique of design methods in engineering – it is an effort to describe the extent to which the engineering profession is invested in conventionalism and empiricism, whether consciously or not.

The conversion from designs based on allowable stresses to designs based on member strength, using partial safety factors for loads and resistance, presents an example of alternative conventions in structural design. In allowable stress design, the conventional maximum loads on a structure are used to determine the elasticity-based stresses in the elements of the structure. These stresses are then verified to be less than the maximum stress in the material modified by a safety factor. In the other method, now adopted by design codes for most structural materials, the strength of an assembly (such as a reinforced concrete beam) is assessed, and partial safety factors are applied to require consideration of greater loads and reduced capacity of the member to achieve a desired level of safety. The existence of two very different ways of designing a member, such as a steel beam or a reinforced concrete column, implies that the design procedure is accepted and applied by the engineering profession as a convention. Examples of the conversion of building codes for reinforced concrete (steel codes underwent a similar conversion) are present both in the US and in the UK. In the US, the American Concrete Institute code for the design of structures in the 1963 edition (ACI, 1963: pp. 318–336) provided alternative methods using working stresses and based on the determination of the ultimate strength of a reinforced concrete beam or column. Subsequent editions of the code discarded the working stress method and embraced the ultimate strength method. A similar change in design procedures was made in the UK when the British Standard 8110 was modified to be based on strength design. The design of steel structures underwent similar modifications in the 1980s and 1990s, changing from an allowable stress-based criterion of acceptability to an acceptance criterion based on ultimate forces and moments. The characteristics of the three design methods for steel, concrete and wood are given in Table 5.1.

The presence of very different methods of analysis that depend on very different assumptions and procedures suggests that the design of steel, reinforced concrete, wood or masonry structures is based largely on convention.

Geschwindner (1988) describes the merits and demerits of LRFD in steel at a very early stage in the adoption of this method. Among the advantages he notes the application of consistent reliability to different members, and even to different materials as concrete, wood and masonry adopt LRFD approaches. He also notes the complexity of the new method and its basis in statistics that may be an obstacle to its adoption. By the present time, LRFD is the generally

Table 5.1 Comparison of analysis and design methods

	Design designation	Loads	Analysis	Resistance
Steel	Working stress	Service loads	Elastic	Linearly elastic
	Load and resistance factor	Factored	Elastic	Plastic
	Limit states	Partial load factors	Design factors, combination factors	Plastic
	Allowable strength	Factored	Elastic	Plastic with safety factor
Concrete	Working stress	Service loads	Elastic	Elastic cracked section
	Strength	Factored	Elastic	Inelastic ultimate resistance
	Limit states	Partial load factor	Design factors, combination factors	Inelastic ultimate resistance
Wood	Working stress	Service loads	Linearly elastic	Linearly elastic
	Load and resistance factor	Factored	Linearly elastic	Linearly elastic
	Limit states	Partial load factor	Design factors, combination factors	Linearly elastic (US NDS) Test-based (Eurocode)

accepted basis for the design of steel structures. However, AISC has also introduced an allowable strength design method as an alternative to LRFD. This method bears superficial similarities to allowable stress design, while introducing another conventional treatment of steel design. The features of this method are also shown in Table 5.1. Similarly, between 1990 and 1995 Eurocode 5 (BSI, 2004b) transitioned from a permissible stress basis to a full limit states design basis. This conversion was more consistent and completed more quickly than in the US.

In the US, the ultimate strength design method developed for reinforced concrete in 1963 was not conceptually a load and resistance factor method, but it has many of the characteristics of such a method including load and undercapacity factors. The LRFD method for steel structures was introduced in the 1990s, and a unified LRFD/allowable strength design method appeared in 2005. A strength design alternative was written into the masonry code in 2002, where it remains as an alternative method. The recent conversion of the US *National Design Specification* for wood (AWC, 2018) is described later in this chapter.

In the Eurocodes, the newer design method is referred to as ‘limit states’ design. In this procedure, which in general outline is similar to ‘load and resistance factor design’, service loads are increased by a statistically determined multiplier to a presumed ultimate loading, while resistance is similarly modified by a statistically determined reduction factor. The comparison of the increased load and the reduced resistance assures an acceptable level of structural safety.

Eurocode: Basis of structural design (BSI, 2002a) uses a wind load calibrated based on its variability, modified by a standard γ-factor (>1.0) to account for the variability in loading and a further ψ-factor (<1.0) to account for the unlikelihood of maximum loading of two types simultaneously. Unlike the US counterpart, the European standards maintain a consistent basis for determination of design loads across the different types of structural loading (i.e. live load, wind load, seismic load) (see below for modification of partial safety factors for loading in ASCE 7).

Among the reasons advanced for the overall changes to LRFD or LSD was to separate the uncertainty of loading: wind, earthquake and gravity, from the uncertainty of the strength of engineering materials. This would have allowed a consistent reliability across all materials and all forms of loading. This idea of consistency in the application of uncertainty to loads and to the resistance of materials was central to the adoption of the newer form of design. In the earlier history of the adoption of this method in the US, load factors were proposed and used for wind loads and earthquake loads: these load factors were thought to reflect the uncertainty in the calculation of these loads, in the understanding of these loads and in the estimation of their recurrence. However, in the 1993 version of ASCE 7, explicit load factors for seismic loads were abandoned. The uncertainty in these loads is now incorporated into the procedures for estimating the loads, and the load factor was modified to 1.0. In the 2010 edition, the load factor for wind loads was also reset to 1.0.

5.2. Material design specifications

In Eurocode 5 (BSI, 2004b), which covers the design of wood structural elements, the conversion from 'permissible stress' to limit states design took place between 1990 and 1995. In the limit state procedure, the ultimate loads (based on factors applied to structural actions) are compared with the limiting condition of a structure (e.g. rupture). In Eurocode 5, wood species are organised into classes (e.g. C16, D30), depending on their strength and stiffness. The characteristic stresses given for the various classes are strength values. This is a different procedure from that given in the *National Design Specification* (AWC, 2018), as is discussed below.

The most widely circulated code for the application of wood products in the US is the *National Design Specification* (NDS). Because wood is a naturally occurring material, well known to have significant variations and significant defects, it is a material that is difficult to reduce to readily applicable conventions. The basis of the NDS is the determination of allowable stresses in wood. These stresses are based on species groups, different species of tree supposed to have similar properties, and grades, determined either by visual inspection or by an automated mechanical process known as machine stress rating. The allowable stresses depend mostly on detailed measurement of the size of the knots, the density of the grain and other visible defects. Examples of a set of such rules can be obtained from the published literature (WWPA, 2021).

Visual grading itself is also evidence of the presence of empirical design. Whereas the application of such exacting rules as the determination of the dimensions and frequencies of knotholes, the maximum and average dimension, grain spacing and slope are called for by the grading rules, experienced lumber graders can determine at a glance the probable grade of a batch of lumber.

For timber structures according to Eurocode 5 (BSI, 2004b), the design value of the resistance is given as

$$R_d = k_{mod} \frac{R_k}{\gamma_m} \quad (5.1)$$

where R_d is design resistance, k_{mod} is a modification factor, R_k is the characteristic value of load-carrying capacity and γ_m is a partial safety factor. This displays the general limit states approach of Eurocode 5.

However, the conversion of the NDS for wood structures from allowable stress design to LRFD design shows some of the inconsistencies in this conversion. The compilation of the existing tables for allowable stresses in wood species groups are a convention that the wood industry has used for a very long time. This information has not changed substantially in the adoption of an LRFD format. Base allowable stresses in wood are measured and collected in the same way as they were in allowable stress design. However, as the loads in design are increased by load factors, it is necessary to reduce the load effect when considering the strength of the materials being designed. This is done through a 'format conversion factor', which is a constant used to reduce the bending moment or shear on a wood member so that the reduced values can be used to estimate the stress to compare to the stress tables that are unchanged from their use in allowable stress design. The format conversion factor differs for differing types of structural members and reflects the basis of the LRFD for wood in reliability-based concepts, but it is incorporated without an explicit partial safety factor.

Other conventions exist in the code for the design of reinforced concrete structures, ACI 318, (ACI, 2019). The linearisation of the fundamentally parabolic stress–strain law for concrete has been discussed above. An important variable in analysing reinforced concrete beams and columns is the maximum usable strain, fixed at 0.003, in spite of the very wide variation in the measured values of this constant. A different value, 0.0035, is used in Eurocode 3 (BSI, 2005). Similarly, the elastic modulus of concrete, although highly variable in practice, is taken as $4700\sqrt{f'_c}$ by convention (ACI 318:19.2.2.1(b)) (ACI, 2019). In Eurocode 2 (BSI, 2004a), the value of the modulus of elasticity is simply prescribed on the basis of the strength class of the material. For comparison, 28 MPa concrete is assigned a modulus of elasticity of 25 GPa in ACI 318 and 33 GPa in Eurocode 2. Some limitations of the application of the Bernoulli–Euler theory of bending to reinforced concrete are considered below; however, it is, by convention, the basis of the flexural design of reinforced concrete beams and columns.

An important convention in both the steel and the concrete codes is the general use of elastic analysis, even though the design strength of the material is associated with considerations of plasticity. This applies to both steel and concrete structures. When a structure responds plastically to bending moments, by developing a hinge at a plastic moment level, M_p, it is possible to describe a means of analysis of a structure based on this plastic behaviour. Elastic structural theory does not really explain the distribution of moments where a hinge has formed or the behaviour of a structure after hinges have formed. Horne (1971), among others, describes means of analysing such a structure based on the theory of plasticity, which may be consistent with the adoption of a plastic approach to the strength of the beams and columns within the structure. In concrete design, small redistributions of support moments may be allowable (both in ACI 318 and in Eurocode 2), based on a limited application of plastic analysis.

The coefficients used in the calculation of shear and bending moment in beams and in calculating moments in two-way slabs have persisted through many editions of ACI 318 despite their being very rough approximations of the actual conditions. It is evidence for these bending moments resulting from conventions, rather than detailed analysis, that the same method of calculating bending moments in regular beams has been used for over 70 years, while the ability to calculate structures more accurately has increased greatly. The call for simplicity made by Beeby and Taylor (1978) appears to be heeded in this case, at least. On the other hand, the use of such coefficients, begun in very early editions of ACI 318 for two-way systems, has recently been omitted from ACI 318-19.

5.3. Fundamental engineering conventions

In the practice of structural engineering, some conventions are recognisable, particularly in the case of building codes that specify loading or methods of construction, or material codes that prescribe the ways that reinforced concrete or steel are to be applied as elements of a building structure. Beyond the building codes and material codes, the profession has adopted other implicit conventions as a means of simplifying the design of a building structure. These conventions include safety factors, conventional loads, sizing upwards and theories of structural mechanics.

5.4. Conventional loads

Dead loads, for known quantities of known materials, may be calculated on a straightforward volumetric basis. However, the application of dead loads often makes use of conventions. A large proportion of the dead loads in a building structure may be included in a lump sum allowance, called 'superimposed dead load', which varies from 0.25 to 0.5 kN/m^2. For a reinforced concrete slab with drop panels, a thickened slab found only adjacent to the columns, engineers rarely distribute the varying floor dead loads according to their location. Instead, the dead load of the drop panel is distributed over the entire bay, or the larger load is assumed for the entire bay. For one reason, the analysis methods customarily used for reinforced concrete often do not support the use of two different floor loads. For another, the use of two different loads in different parts of the floor results in tedious calculations. Structure self-weight loading in steel structures is rarely calculated directly, member by member. Instead, a conventional value of the average weight of the steel structure in N/m^2 is added to the floor loads. Typical estimates of this quantity range from 0.30 to 0.50 kN/m^2. These time-saving measures can be justified by an empiricist appeal, as described in Chapter 3 and below.

Many of the live loads that appear in building codes are more the result of convention than of any direct analysis of building live loads. The commonly used live load for residential structures of 2.00 kN/m^2 appears in textbooks from the nineteenth century onwards (e.g. Hatfield, 1871), without any discussion of a rational determination of the live load. In the building codes in the US for an office building, a live load of 4.0 kN/m^2 is required for corridors above the first floor in an office building. For the occupied areas of an office building, only 2.50 kN/m^2 is required; however, an additional 750 N/m^2 allowance is made for partitions, on the grounds that partitions may be modified over the life of the building. The total loading of 3.25 kN/m^2 is often simplified to 4.00 kN/m^2 for the entire floor, because the combined office floor loading is similar to the corridor loading, and it is not possible to predict where the corridors will be located for the entire life of the building. It is also significant that the design of the supporting structure of the floor usually considers a uniformly distributed load on the entire floor, making it difficult to manage the difference in total live load between the offices and the corridors.

In Eurocode 1 (BSI, 2002b), the uniformly distributed loading for office buildings is 2.00–3.00 kN/m^2, the same range as in the US. The imposed load in corridors (category C3) is 3.00–5.00 kN/m^2. In the UK National Annexe, the partition loading varies with increasing area, but, on average, is similar to US practice with a value of approximately 0.80 kN/m^2. The previous discussion of the widely observed practice of designing office spaces for corridor loads is similarly advantageous in ordinary office buildings in the UK.

5.5. Factors of safety

The idea of safety factors is present in the thinking of every contemporary structural engineer. It is considered in nearly every case of engineering design, even when the codes do not call for an explicit safety factor. In general, the safety factor is a ratio of measured strength to allowable stress – that is, for steel that has an average yield stress of 0.25 MPa, a safety factor of 1.67 gives an allowable stress of approximately 0.15 MPa. Modern codes that have discarded the use of a single safety factor are nevertheless calibrated to have the same reliability as the population of structures designed according to the traditional single safety factor. This is the case, for example, according to the discussion of Galambos and Ravindra (1978) on the calibration of an LRFD code for steel structures. These authors further note that this method of calibrating an LRFD code 'has the advantage of utilising past experience'.

Most of the safety factors in widespread use have decreased over the previous century and a half. Where safety factors of 4 or 5 were widely used for structures of wrought iron or steel, the modern safety factor for this material is less than 2. Whereas safety factors of 3 to 5 were prescribed for wood in the nineteenth century (Kidder, 1886), the implied safety factor according to the current method of evaluating allowable stress in wood is approximately 2 (ASTM, 2022), although this safety factor is applied to the strength of a low percentile group of tested specimens. For metal structures this change in safety factor can be attributed to improved manufacture, but no such claim can be made for changes in the safety factor for wood. Other safety factors in widespread use include a safety factor of 1.5 for overturning of a structure due to wind pressure or soil pressure, 3 for determining the allowable stress in soil, a safety factor of 2.5 for reinforced concrete, and a safety factor of 3 or 6 for masonry, depending on the level of inspection the work is subjected to. For many tension members, such as steel cables, a larger safety factor of around 5 is used (The Engineering ToolBox, 2021). If there is no specific guidance on what factor of safety to use, an engineer who has access to testing data will usually choose a factor of 2 or 3, or will choose a larger safety factor if a significant body of test results is not available. The use of such safety factors may be more conventional than rational, as the magnitude of the safety factor often appears to be arbitrary.

The very wide range in these safety factors and the difference between the numbers used in previous years and the contemporary safety factors may be taken as evidence that these numbers are not always determined rationally. Beyond a rational basis, they also include a large measure of historical continuity, intuition, favourable or adverse experience and following of conventions. The result is a combination of reliance on experience and reliance on conventions in the determination of appropriate safety factors for materials and assemblies.

The argument that preceded the adoption of load and resistance factor design methods is that these methods were rationalising the design of structures and eliminating the application of arbitrary safety factors in design. An extensive discussion of the development of target reliability indices for LRFD design of bridges indicates that the reliability of previous bridges was used as the basis for the reliability of new bridges (Taly, 2014). In another discussion on LRFD for buildings, it is unclear that there is a rational basis for the 1.2 partial safety factor used in the US for dead loads (Heger, 1994). All the same, the effort over the past two or three decades has been to harmonise the safety factors across the range of structural materials used in practice.

5.6. Conventions of structural mechanics

The idea of empiricist appeals has been introduced previously in Chapter 3. In the following, we describe some inconsistencies that are embedded into the design procedures used by contemporary engineers. These inconsistencies or exceptions to rational analysis are generally justifiable by empiricist appeal: that structures that have been designed by these methods have been found to function acceptably.

Although theories of structural mechanics find a use in material codes, notably the codes for wood, reinforced concrete and steel, the conventions that underlie these theories cannot necessarily be ascribed to the code but represent canons that are taught to every emerging structural engineer and are accepted generally by practising structural engineers. Such conventions are discussed here.

The theory of elasticity – that is, the notion that stress in a material is proportional to the strain – is a computational convenience that is used in the construction of most of the theories of structural behaviour. The only structural material for which this type of behaviour is really evident is steel below the yield stress, where stress–strain plots appear to be perfectly linear. The character of the relation between stress and strain is much less straightforward for wood. Figure 5.1 shows the stress–strain curve of a wood specimen in tension parallel to the grain, especially the departure from linearity at higher load levels.

The application of stress–strain laws to concrete structures is similar. The stress–strain curve of concrete in compression is approximately parabolic (Figure 5.2), but for purposes of analysis is assumed to be linear. In testing of concrete specimens, the modulus of elasticity of the concrete exhibits very wide variations. However, for deflection calculations of reinforced concrete structures, a single prescribed value of the modulus of elasticity is used. Moreover, in deflection calculations, it is exceptionally difficult to determine where the concrete is cracked, and what is the depth of the crack penetration, so a conventional, approximate formulation is used for the bending stiffness of a reinforced concrete beam (ACI, 2019).

Figure 5.1 Stress–strain curve for wood, illustrating non-linearities along the grain and perpendicular to the grain (Figure by Daniel Hopkin, adapted from Buchanan (2001))

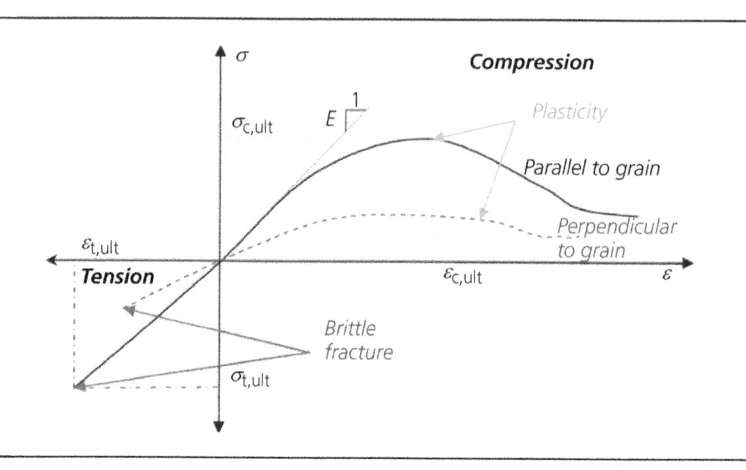

Figure 5.2 Concrete stress–strain curve (Figure by Si Shen, https://www.si-eng.org/post/a-step-by-step-anatomy-of-concrete-stress-strain-curve-1)

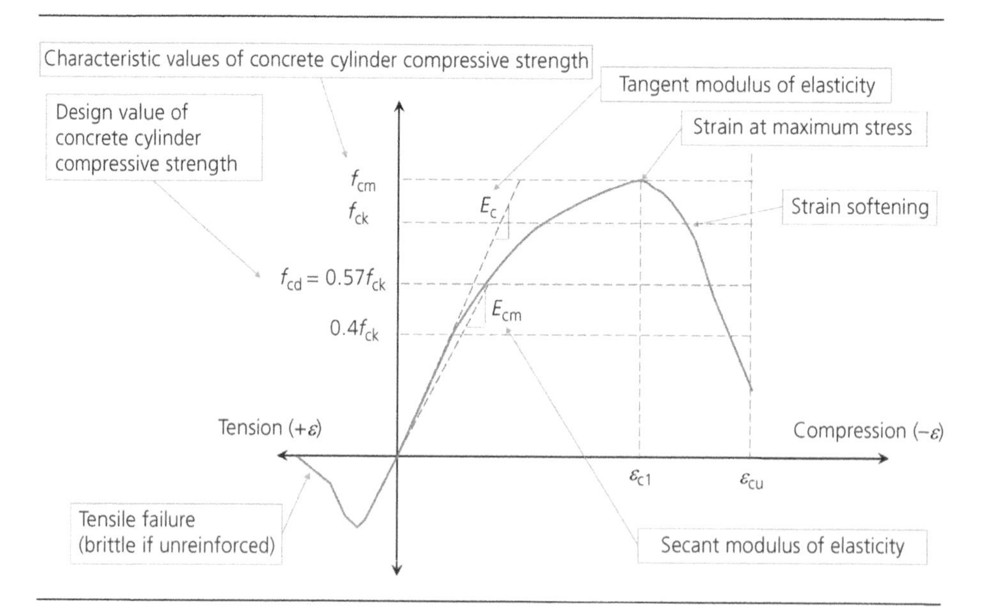

Similarly, the behaviour of unreinforced and reinforced masonry is not only linearly elastic by convention, with the stiffness depending on the load level and with large variations in the initial modulus of elasticity, which result from confounding factors such as variability in the concrete material, variations in the cracking of the masonry, relative stiffness of mortar joints and masonry units and other factors. However, this assumption of linear elasticity is used to determine the relations between the bending moment, stress in the masonry and stress in the steel. These quantities are then applied to the determination of the capacity of the masonry.

The behaviour of masonry structures can be very complicated. The usual form of a reinforced masonry beam is in a wall subjected to two-way bending. The combination of uncertain effects in the 'plate' and the issues of understanding strain in a cracked section make the application of this theory of bending extremely problematic. However, standard textbooks and industry design aids for the design of reinforced masonry walls use this general theory of bending, applied to a plate spanning vertically for application to the design issues in a reinforced masonry wall.

Accepting linearly elastic behaviour for materials such as wood and concrete is contrary to observable fact but can again be understood as one of the conventions of structural engineering. The justification of the use of incomplete theories of material behaviour is an empiricist appeal: it is possible to construct a theory based on linearly elastic wood or concrete which allows the sizing of beams and columns that are unlikely to fail. It is important to note that, in spite of the evidence presented here that the stress–strain behaviour of wood and concrete is not linear elasticity, nevertheless the engineering profession is comfortable with the application of a linear stress–strain law to these materials. There is nothing incorrect about this practice: the engineering profession's success in designing reinforced concrete structures testifies to the effectiveness of this convention. It is, however, necessary to accept the notion of linear elasticity in concrete as a convention adopted by the engineering profession and accepted by empiricist appeal.

The design of steel beams often includes interaction with a portion of the concrete slab that the beam supports, known as composite design. Some of the aspects of this form of design are partly conventional, beginning with the selection of the width of the slab that is considered to participate compositely with the steel beam. There is also a theory of 'partially composite behaviour' for which the difficulties in analysis do not normally enter into the design of such structures. The deflections of such beams are calculated solely by lower and upper bounds on the stiffness of a partially composite beam and have only an informal rational justification.

One of the most widely dispersed and widely used notions of structural mechanics is the theory of flexure, or bending, of beams, known as the Bernoulli–Euler theory of bending (Heyman, 1998). This theory holds that the deformation of a beam in bending is such that sections of the beam that were plane prior to bending remain plane after bending. The result of this hypothesis is that there is a uniform gradient of strain within the beam, with an unstressed neutral axis at the geometric centroid of the beam cross-section. This results in a uniform stress gradient, ranging from maximum compression at the inside of the bent curve of the beam to a maximum tension at the other face of the beam. This theory was primarily developed by experiments on wood beams and holds reasonably well for solid rectangular beams made from a homogeneous material. However, it is uncertain that this idealised description of bending is appropriate for reinforced concrete or masonry beams, or for composite steel/concrete beams. The concrete or masonry is expected to crack through the tension zone, and the assumption or even the measurement of the effect of curvature at any given location is a very difficult matter. In a reinforced concrete beam, the cracks in the tension zone concentrate the deformations in such a way that it is very difficult to speak of 'strain' except in an averaged sense. Where the results of such analysis speak of concrete strain in the tension zone, they are referring to a conventionalised strain that cannot be measured.

5.7. Columns

The Euler buckling theory is extensively applied to wood and steel columns. The maximum design stresses are modified for the failure of the material and for inelastic buckling at lower slenderness ratios. The ideas behind this theory are used in the steel, concrete and wood design codes: in the steel and wood codes, the influence of larger slenderness ratios is taken into account, while in the concrete code, the influence of this theory on the behaviour of beam-columns is considered. The Euler theory of buckling, as applied in modern practice, is developed for a simply supported, perfectly straight elastic column with perfectly concentric loading. It is possible to derive a formula that defines the load at which the column may assume a buckled configuration at a critical load. In this theory, the capacity of the column is not diminished up to the critical load.

Under practical conditions, the result is quite different and this theory must be modified. The first modification is to truncate the column buckling curve at a limiting stress for the material of the column. The second modification is to find an empirically determined interpolation curve to mediate between the elastic buckling behaviour at high slenderness ratios and the inelastic behaviour at lower ratios. These two curves are shown schematically for steel and for wood in Figures 5.3(a) and 5.3(b).

Figure 5.3 Column design curves in steel and wood: (a) column strength curve in steel (based on AISC, 2017); (b) column strength curve in wood (based on AWC, 2018)

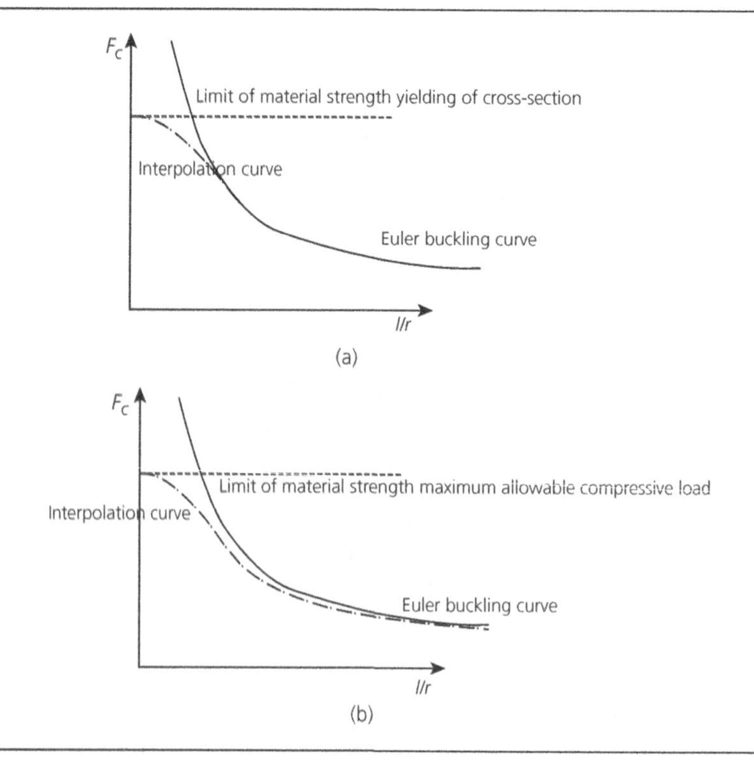

In the nineteenth century, column behaviour was addressed differently: a theory based on the initial imperfections was more widely distributed (the Rankine–Gordon theory is more fully discussed in Boothby (2015)). The profession in general was skeptical of the Euler theory as a basis for column design. In advocating a straight-line column capacity, Palmer (1916) declared that

> Experimental results show that the conditions upon which Euler's formula depends only exist where the slenderness ratio is greater than 200.... Since the building codes which regulate modern practice do not permit the use of building columns in which the slenderness ratio exceeds values of 120 to 150, Euler's formula is not applicable in practice.

For modern steel UC shapes, the limiting slenderness ratio for the validity of the Euler formula is approximately 113, which represents, for instance, a 203 × 203 × 46 section with an unsupported length of length of 5.8 m, a slenderness that is rarely seen in practice.

5.8. Beam-columns

Beam-columns – that is, beams that are also subjected to axial force or columns that are also subjected to bending – are considered in steel design by a simple, empirical straight-line interaction formula, as the mechanics of this situation – combined compressive and bending stress plus the influence of buckling – are too complicated to be considered analytically. In concrete design, the procedure is more rational, but must be handled inversely, with an interaction relationship developed for each given column and its reinforcement and then tested (by trial and error) against the properties of the proposed section. The difficulties of this inverse procedure often lead to the use of simplified or empirical methods for sizing reinforced concrete beam-columns.

5.9. Connections in wood and steel

Steel connections, like the remainder of steel structures, are designed by a combination of rational and empirical procedures. Simple connections in steel consist of a pair of angles on either side of the web of the connected member, secured by bolts through clip angles to the web of the connected and the supporting members. Although the rational design of such a member is possible, it is more typical to attribute a level of strength to each row of bolts and to maintain sufficient edge and end distance – these quantities are determined empirically but incorporated within the design specification.

Welded full moment connections in steel are composed more than designed. To ensure moment capacity, the flanges are welded to a supporting structure, to an end plate or to another beam using a full penetration groove weld. A full penetration weld is not sized: its size is equivalent to the flange thickness of the beam in the connection.

The *Steel Construction Manual* (AISC, 2017) also shows some standard connections that are preapproved on the basis of prior experience. The welded connections include 'prequalified welds' that means that the procedures shown have been proved by prior experience to be effective. This can be understood as a further application of empirical design, as experience is the primary guide in the effectiveness of these connections.

Wood connections function by a wide variety of empirical rules for size, spacing, edge distance and end distance. These rules, codified in the US in the *National Design Specification* (AWC, 2018), attempt to cover the variety of available fasteners, bolts, lag bolts, screws, dowels, drift pins, nails, split ring connectors and so on and special conditions such as the density of the wood, the use of metal side plates, grain orientation and other factors. As an example, the capacity of 12 mm dia. bolts in double shear in spruce-pine-fir (specific gravity = 0.42) is given in the NDS as 4 kN parallel to the grain, modified for the use of metal side plates, force in a different direction with respect to the grain, for conditions of single or double shear and so on. The minimum end distance for full strength in tension is 7 bolt diameters. Minimum spacing along the force line is 4 bolt diameters. Almost all of the information in the above requirements is empirical. The basic bolt design value is surely an interpolation of limited testing information, while the strength increase due to metal side plates, and the edge distance, and distance and spacing are simply the result of favourable experience with connections that are proportioned in this fashion. For the reader with sufficient interest, a three-page-long design example of a bolted connection in wood is given in Showalter (2016). A critical reading of a code, such as the NDS, shows significant empirical content, and the parts of the code that deal with connection design have a large proportion of empirical information. Although it is possible to use reason and theories of mechanics to estimate the shear capacity of a nail, such reasoning would likely yield incorrect results. In any event, the results would be biased by nail spacing, end and edge distance, differences in the properties of the wood that holds the nails, the influence of out-of-plane loading, and a variety of other factors that make the rational calculation of the capacity of a nail a very difficult exercise. Instead, the tables are developed in a semi-empirical fashion. It is considered that the capacity of the nail will be increased roughly in proportion to the diameter of the nail, and a few tests are done under laboratory conditions to arrive at initial estimates. A conservative safety factor is applied (experience shows that a larger safety factor is appropriate for connections due to the number of unknown factors and the critical conditions induced by a connection failure) and the estimates based on experience, tempered by some rational justification, are compiled into a table of fastener capacities.

The elastic analysis of in-plane eccentric connections is cited in codes on both sides of the Atlantic, including Eurocode 5: 8.3 (BSI, 2004b) and the AISC *Steel Construction Manual*: 10-131 (AISC, 2017).

As pointed out in Chapter 3, there is a body of research that discredits this point of view. At least in part, the continuation of this form of analysis is a combined product of experience and comfort with this form of analysis and, again, the empiricist outlook that it has worked previously. For wood structures, the evidence in favour or against this type of analysis is less certain, due to confounding effects of grain direction, fastener slip and other effects common to wood structures, but the simple eccentric analysis in Eurocode 5 may well be an oversimplification. In the US, the NDS (11.1.3) simply says 'don't use this type of connection' (AWC, 2018).

5.10. Conclusions

Although the design of steel, concrete, wood and masonry structures often has a basis in rational principles, the design specifications and the detailing of such structures has elements of empiricism or relies on empiricist appeals. There are two types of empirical design considered in the foregoing: one is the simple re-use of elements or assemblies that have been proved to work in the past and are repeated in the next design. The other is the adoption by the profession or its governing bodies of conventionalised procedures that may result from reason or from experience and have continuing applicability by convention. These conventions are embodied in building codes or as part of the traditional basis of engineering. Some of these conventions can be seen to come from reasoned consideration of natural phenomena involving structures, while others do not appear to be rational on close inspection. It is important for a designer to understand the extent to which conventions, rather than the application of reason or of scientific investigation, may be determining the standards on which they are basing their design activities.

REFERENCES

AISC (American Institute of Steel Construction) (2017) *Steel Construction Manual*, 15th edn. AISC, Chicago, IL, USA.

ACI (American Concrete Institute) (1963) ACI 318-63: Building code requirements for structural concrete. ACI, Detroit, MI, USA.

ACI (2019) ACI 318-19: Building code requirements for structural concrete. ACI, Farmington Hills, MI, USA.

ASCE (American Society of Civil Engineers) (2016) ASCE 7-16: Minimum design loads and associated criteria for buildings and other structures. ASCE, Reston, VA, USA.

ASTM (2022) ASTM D245-22: Standard practice for establishing structural grades and related allowable properties for visually graded lumber. ASTM International, West, Conshohocken, PA, USA.

AWC (American Wood Council) (2018) *National Design Specification*. AWC, Leesburg, VA, USA.

Beeby A and Taylor H (1978) Use of simplified methods in CP110 – is rigour necessary? *The Structural Engineer* **56(8)**: 209.

Boothby T (2015) *Engineering Iron and Stone: Understanding Structural Analysis and Design Methods of the Late 19th Century*. ASCE Press, Reston, VA, USA.

BSI (2002a) BS EN 1990:2002 Eurocode: Basis of structural design. BSI, London, UK.

BSI (2002b) BS EN 1991-1-1:2002: Eurocode 1: Actions on structures – Part 1–1: General actions – Densities, self-weight, imposed loads for buildings. BSI, London, UK.

BSI (2004a) BS EN 1992-1-1:2004: Eurocode 2: Design of concrete structures – Part 1–1: General rules and rules for buildings. BSI, London, UK.

BSI (2004b) BS EN 1995-1-1:2004: Eurocode 5: Design of timber structures – Part 1–1: General – Common rules and rules for buildings. BSI, London, UK.

BSI (2005) BS EN 1993-1-1:2005: Eurocode 3: Design of steel structures – Part 1–1: General rules and rules for buildings. BSI, London, UK.

Buchanan AH (2001) *Structural Design for Fire Safety*, 1st edn. Wiley, Chichester, UK.

Galambos TV and Ravindra M (1978) Load and resistance factor design for steel. *ASCE Journal of the Structural Division* **104(9)**: 1337–1353.

Geschwindner L (1988) LRFD and the structural engineering curriculum. *Engineering Journal* **25(2)**: 79–84.

Hatfield RG (1871) *The American House-Carpenter*, 7th edn. Wiley, New York, NY, USA.

Heger F (1994) Discussion of calibration of current factors in LRFD for steel. *Journal of Structural Engineering* **120(9)**: 346–347.

Heyman J (1998) *Structural Analysis: A Historical Approach*. Cambridge University Press, Cambridge, UK.

Horne MR (1971) *Plastic Theory of Structures*. MIT Press, Cambridge, MA, USA.

Kidder FE (1886) *The Architects' and Builders' Pocket-book*, 3rd edn. Wiley, New York, NY, USA.

Palmer C (1916) *The Constructive Design of Modern High Buildings*. MS thesis, Pennsylvania State University, PA, USA.

Runes D (1983) *Dictionary of Philosophy*. Philosophical Library, New York, NY, USA.

Showalter J (2016) Design of bolted connections per the 2015 NDS. *Structure*. November. https://www.structuremag.org/wp-content/uploads/2016/10/C-PracSolutions-Showalter-Nov16-1.pdf (accessed 29/05/2023).

Taly N (2014) *Highway Bridge Superstructure Engineering*. CRC Press, Boca Raton, FL, USA.

The Engineering ToolBox (2021) *Wire Rope – Strength*. https://www.engineeringtoolbox.com/wire-rope-strength-d_1518.html (accessed 08/05/2021).

WWPA (Western Wood Products Association) (2021) *Western Lumber Grading Rules*. WWPA, Portland, OR, USA.

Boothby T
ISBN 978-0-7277-6633-5
https://doi.org/10.1680/edse.66335.063

Chapter 6
The ethical use of empirical design

6.1. Empirical design as an engineering method

In reviewing the ethical standards of engineering societies, it is possible to infer an outlook on non-standardised design methods, such as empirical design. In the US, the National Society of Professional Engineers (NSPE, 2023a) maintains a website with a thorough description of ethics cases that have been brought to their board of ethical review, along with the findings of the board. In addition to the fundamental canons of ethics, NSPE presents rules of practice, among which is (II.1.b): 'Engineers shall approve only those engineering documents that are in conformity with applicable standards.' In some of the circumstances outlined in this chapter, this may present an ethical challenge to the use of empirical design.

In the UK, the governing body is a collaboration between the Engineering Council and the Royal Academy of Engineering. Together, they publish a statement of ethical principles for engineers, which is available online (Engineering Council, 2023). The portion of this statement most applicable to the present topic is under the heading 'Accuracy and rigour': 'Engineers should… present and review theory, evidence and interpretation honestly, accurately, objectively and without bias, while respecting reasoned alternative views.' Empirical design can be considered a reasoned alternative view – not literally 'reasoned', but 'resulting from experience'. The important philosophical distinction between these two ideas was presented in Chapter 2. Also, in Chapter 5, we have noted instances in which published standards may be erroneous, inconsistent, subject to erroneous interpretations, and not always meeting the standard of 'rigour'.

It appears, however, that the boards that review ethics for engineers both in the UK and in the US suggest that there is an ethical duty to conform to existing standards. NSPE Board of Ethical Review Case No. 02-5 (NSPE, 2023b) touches lightly on this topic. In this case, an engineer who followed previous guidance in the determination of snow loads instead of guidance published more recently (but not in the form of a standard) was found to be practising ethically. This at least establishes the principle that an engineer can practice ethically, while using an obsolete method to design a structure. However, it is clear enough from the rules of practice (particularly II.1.b, quoted above) that an engineer has an obligation to follow standards when available.

In spite of the presence of empirical design, empirical knowledge and empiricist corrections in the codes of practice of structural engineering, explicit empirical design methods are not generally available in codes of practice. There are, or were, a few exceptions. In TMS 402 (TMS, 2016), the code of practice in the US for the design of masonry structures, an empirical method was available for use in limited circumstances – moderate seismic zone, moderate wind velocities, heights four storeys or fewer. This method relied on height/thickness ratios for the walls and a simplified calculation of compressive stress on the gross area of masonry for the design of both load-bearing and non-load-bearing walls. However, this method has recently been withdrawn from this standard. The British Standard DMRB: CS 454 is used worldwide for the assessment of masonry bridges (Highways England, 2019). This procedure contains a preponderance of empirical provisions, especially in the use of condition factors for the strength of an existing bridge. The condition factors call for significant reductions in calculated strength based on visual observation.

Another case of an empirical procedure is the use of engineering judgement in earlier editions of the *Manual for the Condition Evaluation of Bridges* (AASHTO, 1994), re-edited as the *Manual for Bridge Evaluation* (AASHTO, 2008), which allows load testing in limited circumstances as the only admissible empirical procedure. The earlier *Manual for the Condition Evaluation of Bridges* contains the following passages.

> 6.5.5 For redundant bridges, where necessary details, such as reinforcement in a concrete bridge, are not available from plans and field measurements, a physical inspection of the bridge by a qualified inspector and evaluation by a qualified engineer may be sufficient to determine Inventory and Operating ratings.
>
> 7.7 The evaluation of highly unusual structures and special conditions requires good engineering judgment. In the load-rating of such structures, the original design method may be used, with adjustments to reflect the actual condition of the structure.

These passages, although obsolete, are modern in origin and reflect sympathy to the use of original design methods, including empirical design, and the use of empirical evidence of a bridge's safety. Even though this assessment method no longer exists in the current versions of this code, it is still widely used, according to the national bridge inventory (Federal Highway Administration, 2022). Data on masonry bridge assessment in this document showed that load ratings for most of the masonry bridges were determined by 'engineering judgement'. The tension between engineering judgement, which it is a licensed professional's privilege and duty to exercise, and the canonised ethical obligation to follow accepted standards is not often addressed in discussions of engineering ethics.

There are four primary areas where an engineer may choose to apply empirical design. Each of these areas has a different category of ethical involvement.

6.1.1 Empirical design used to substantiate the findings of an analytical investigation

The idea of checking your own work or having your work checked is fundamental to the ideas of engineering ethics promoted by engineering societies. The merits of empirical design as a method for testing the accuracy of calculations and its ease of application are evident. As the application of empirical design depends only on experience, it is easy enough to say, 'have I seen something like this before?' and to apply this experience to determine the adequacy of the design under consideration. The empirical estimation of beam sizes can easily be applied to choices of beams in an engineering project. The number of bolts in a connection can be remembered from previous experience or reviewed in an existing calculation. For more complicated or system-wide calculations, empirical design also applies. For the sizing of shear walls in an earthquake-resistant structure, the empirical rules give the cumulative length of shear walls in a building as a proportion of the perpendicular length of the building, and the number of storeys (see Boothby, 2018: Table 10.2) may not be suitable for final design, but they provide a relatively simple check, reinforced by experience, on this design question.

Other empirical formulas serve a similar function. For instance, the span–depth ratios for a beam, height–width ratios for a wall and height–width ratios for a column give an immediate check on the results of analysis, that either verify a design that has been completed, or give a range for a design that is being undertaken.

Of course, if the sizing and reinforcement of structural members are completed by a computer programme, it is necessary to verify the output to make sure that the design remains within parameters established by experience. Beams that exceed recommended span–depth ratios or overly slender columns may indicate errors in load input or errors in selecting items to check (such as deflections). A robust set of empirical rules is an asset in completing checks of this nature.

Empirical design is also a resource in error detection, both in design and in the investigation of built works. Experience, direct or distilled through span–depth, width–height and other ratios, provides immediate guidance for the validation of the design or of the configuration of a structure. As error checking and error detection are practices that are an ethical obligation, empirical design can be an effective way of meeting these obligations.

6.1.2 Empirical design accompanying a routine analytical procedure

An engineer who regularly designs four-storey wood-framed apartment buildings has no further need of using further analytical calculations to size interior bearing walls. From one project to the next, the spacing of these walls is standard, approximately 6–7 m, the live load is the same on the floors supported by these walls. As a result of the consistency of this structure type, there is no reason to expect any deviation from the results of a previous analytical design. Thus, the engineer can responsibly call out all such bearing walls the same or standardise the design of these bearing walls. Similarly, the floor joists have spans that are consistent from one project to the next. Although there may be availability problems with one material or the other, a description of the allowable bending stress (AWC, 2018) or the timber class (Eurocode 5 (BSI, 2004)), and a standard size adopted would be possible. This statement applies to many other types of structures, including low fill height retaining walls, in which a thickness and reinforcement of the wall may be determined by practice, site walls that do not need appreciable strength, or other similar features.

6.1.3 Empirical design as a substitute for an analytical procedure

There are simply cases in the practice of engineering, especially in the assessment of existing structures, in which an analytical method does not address the design or evaluation that needs to be performed.

Often, in engineering practice a proprietary system is used for which the manufacturer does not divulge the design details of the system. In such cases, the engineer of record is obligated to insist on certification of the design by the manufacturer of the system, but there are issues of the interface between the given system and the whole building that require better knowledge of the proprietary system than is necessarily available to the engineer of record. In such a case, it is reasonable to appeal to experience, either acquired personally or the experience of others with the proprietary system in question.

In the following Chapter 7, we will discuss the application of empirical design to the assessment of historic structures. That discussion will be more directly on the use of empirical design for performing such assessments. The discussion in the present chapter focuses on the propriety of the use of empirical methods. Returning to some of the philosophical discussion in the early chapters of this work, every engineer puts faith in the evidence of their senses. This faith is in most cases justified, and in many cases is reinforced by analytical engineering calculations. Difficulties may arise when the structure in question is not amenable to analytical calculations. Sometimes the strength value of the material is unavailable, sometimes the form of the structure is too complex for analytical calculations, and sometimes the interior features of the structure are unknown. In all these cases, the loading history and durability of the structure can be taken into account. It is difficult to claim the omission of calculations as a breach of duty when a method for calculations is unavailable or reports values inconsistent with experience. The presence of internal reinforcement may be confirmed by the durability of the structure. As an example, many arch bridges are found structurally deficient based on the dimensions of the arch barrel, as visible from the side of the bridge, whereas nearly all arch bridges have internal haunching extending up a large proportion of the back of the arch.

Empirical design is not intended to overrule the selection of steel beams or reinforcement for a concrete beam under ordinary situations that are covered by the code. The rational analysis that these codes depend on is also a product of experience, and simple proportioning will not produce better designs.

There are other conditions where the design of a structural element is dictated by factors other than structural design. The use of a 115 mm thick normal weight concrete slab, even where the infill beam spacing is less than 2000 mm, is an example of this. An empiricist knows very well that the slab is acceptable for this span and that no further structural design of the slab is necessary.

6.1.4 Empirical design used on its own for the assessment of a structure

Especially in complex situations, in which the designer can appeal to empirical design to make sense out of a complex analytical design, or in those situations in which a rational design procedure really does not exist, empirical design may be used independently. The discussion of Friedman and Abdelfatah (2019) evaluating nineteenth century fireproof floors, described in the following chapter, falls into this category. It is frequently found in the assessment of existing structures that there is no standard procedure covering obsolete design procedures and obsolete materials. This topic is explored in more detail in the following chapter.

The international standard ISO 13822:2010(E): Basis for the design of structures: Assessment of existing structures (ISO, 2010) allows a presumption of safety for structures built according to earlier codes or constructed in accordance with good construction practice when no codes applied. This is conditional on careful inspection of the structure, a review of the load path and a finding of satisfactory performance to date.

Davis (2012) has an extended discussion of the role of judgement in engineering ethics. In this article the author argues that 'engineering ethics' is not like moral ethics or any other form of abstract ethics but consists of the application of standards and rules of practice to the correct practice of engineering. Within this context of engineering practice, Davis describes the training of engineers in applying codes and standards, and finally notes that one of the most important tasks conducted by an engineer is the exercise of judgement, particularly when there are reasons pulling a design in different directions. The exercise of judgement is required to distinguish between these reasons and choose the more plausible, economic or defensible course of action. This discussion has direct bearing on the ethics of choosing to use empirical design over analytical design, or empirical assessment over analytical assessment. It is common, in the practice of engineering, to be confronted with a structure that has no apparent damage, but that the standard calculations show to be deficient. The exercise of judgement would consider the alternative viewpoints, 'the structure is unsafe' or 'the structure has endured for decades without apparent damage and can be considered safe'. The second statement is not an easy judgement to reach. Both the statement of ethical principles and the code of ethics call for adherence to standards. The type of calculation that failed to certify the structure may conform to standards. The exercise of engineers' judgement requires a further investigation, for instance, reviewing the loading history of the structure, verifying that there is no damage at the locations where failure is predicted by calculations, a review of the general state of the structure, an estimate of the consequences of failure of the structure and other factors. Also warranted is an attempt to retrace the load paths of the structure to find an intelligible analytical means of verifying the strength of the structure. An authentic appeal to engineering judgement might, under such circumstances, be considered ethical.

The latter attitude was permissible up to 1995 in the assessment of existing concrete bridges in the US. The AASHTO *Manual for the Condition Evaluation of Bridges* (AASHTO, 1994) up to that point allowed the exercise of 'engineering judgement', allowing the bridge to be posted for a load level that it was carrying without damage. This apparently circular reasoning can be very effective in postponing the replacement of bridges when their internal configuration is not known.

6.2. Example of off-label prescriptions

An important example from another profession is the right of physicians to use prescription medications for 'off-label' purposes. The 'off-label' use of prescription medications by physicians who are licensed to practice medicine and the selection of methods for design by engineers who are licensed to practice structural design present some similarities. It is legal and arguably ethical for a physician, based on their experience and possibly their understanding of pharmacokinetics, to use a prescription medication for a purpose other than the approved uses of the medication. A physician takes this decision seriously but is usually relying on their experience and intuition and may be overriding the results, or especially the lack of firm results in the scientific literature when they decide to use a medication in this manner (see, for example Wong and Kyle, 2006). This is analogous to an engineer designing a structure that conforms in general to the prescriptions of the building code but whose exact proportions or reinforcement are determined by some method other than advanced structural analysis.

6.3. Engineering rigour

We have previously shown, especially in Chapter 5, that the rigour of many engineering calculations is illusory and that calls for engineering rigour overlook this point. What is needed is engineering experience. A critique of the idea of 'rigour' in the design of reinforced concrete structures is provided by Beeby and Taylor (1978). In that article, the authors point out that all calculations in reinforced concrete are based on approximations of variable and complex behaviour. The authors note that for analysis of indeterminate reinforced concrete beams, the range of expected bending moments can be very large and any calls for rigour in such calculations are futile. This notion extends to many other areas of the practice of structural engineering. Finite element analyses of the strength of masonry, including workmanship, additional features such as haunching, proper assessment of material behaviour and exact determinations of load paths can furnish the appearance of precision, but are often no more accurate than a visual assessment by an experienced engineer. The models that are adopted in the first method have embedded assumptions that are not correct or a requirement to insert parameters that are unknown.

6.4. Conclusions

An engineer practising empirical design for the design of a new structure or the assessment of an existing structure may have to review the ethical issues in their choice of the use of this method. The use of empirical design used as an adjunct to design based on contemporary standards is largely ethically justifiable. However, when it is necessary to use empirical design without the support of contemporary standards, due to incomplete coverage of the standards or due to the assessment that the standards are incorrect, an engineer may not be in compliance with existing standards. In the end, individual engineers need to identify their own principles in applying empirical design, based on their confidence in the procedure and their confidence in the alternative analytical procedures that are available. The more the engineer deals with contemporary structures, the more likely it is that empirical design will be useful only as a secondary adjunct to an analytical design.

REFERENCES

AASHTO (American Association of State Highway and Transportation Officials) (1994) *Manual for the Condition Evaluation of Bridges*. AASHTO, Washington, DC, USA.

AASHTO (2008) *Manual for Bridge Evaluation*, 1st edn. AASHTO, Washington, DC, USA.

AWC (American Wood Council) (2018) *National Design Specification*, AWC, Leesburg, VA, USA.

Beeby A and Taylor M (1978) Use of simplified methods in CP110 – is rigour necessary? *The Structural Engineer* **56(8)**: 209.

Boothby T (2018). *Empirical Structural Design for Architects, Engineers and Builders*. ICE Publishing, London, UK.

BSI (2004). BS EN 1995-1-1:2004: Eurocode 5: Design of timber structures – Part 1–1: General – Common rules and rules for buildings. BSI, London, UK.

Davis M (2012) A plea for judgment. *Science and Engineering Ethics* **(18)**: 789–808.

Engineering Council (2023) *Statement of Ethical Principles*. https://www.engc.org.uk/standards-guidance/guidance/statement-of-ethical-principles/ (accessed 22/03/2023).

Federal Highway Administration (2022) *Downloadable national bridge inventory.* https://www.fhwa.dot.gov/bridge/nbi/ascii.cfm (accessed 02/09/2022).

Friedman D and Abdelfatah M (2019) Hidden strength in historic buildings. *Proceedings of the 2019 IABSE Conference.* New York, NY, USA.

Highways England (2019) Design Manual for Roads and Bridges: CS 454 – Assessment of highway bridges and structures. Highways England, London, UK.

ISO (International Standards Organization) (2010) ISO 13822:3010(E): Basis for the design of structures: Assessment of existing structures. ISO, Geneva, Switzerland.

NSPE (National Society of Professional Engineers) (2023a) *Code of Ethics.* www.nspe.org (accessed 22/03/2023).

NSPE (2023b) *Board of Ethical Review Case 02-5 Professional Competence in Current Structural Design.* https://www.nspe.org/sites/default/files/resources/pdfs/Ethics/EthicsResources/EthicsCaseSearch/2002/BER02-5-Approved.pdf (accessed 03/06/2023).

TMS (The Masonry Society) (2016) TMS 402: Building code requirements and specifications for masonry structures. TMS, Boulder, CO, USA.

Wong D and Kyle D (2006) Some ethical considerations for the 'off-label' use of drugs such as Avastin. *British Journal of Ophthalmology* **15 September 2006**: 1218–1219.

Boothby T
ISBN 978-0-7277-6633-5
https://doi.org/10.1680/edse.66335.071

Chapter 7
Case study I: Empirical design in historic structure assessment and preservation

7.1. Introduction

In the previous chapter, we outlined some professional uses of empirical design, with a focus on the design of new facilities. These uses were, roughly, checking a design produced by analytical methods, as a supplement to analytical methods, and as a substitute for analytical methods. These ideas also apply to preservation engineering; however, the scope for the application of empirical design is broader than for new construction. The method by which a heritage structure is designed is usually not part of contemporary practice: a preservation engineer may choose to research the design procedure that was used at the time of construction. This results in a different set of ground rules and implies additional ways in which empirical design may be used in historic preservation

- in developing concepts that are similar to modern analytical procedures but for which strict code compliance according to modern standards is impossible
- in justifying non-standard analysis procedures to explain the durability of a structure that is apparently deficient
- in justifying a structure that has endured, simply by presuming sufficient strength based on longevity.

The application of the third method, as described in Chapter 6, may require that an approximate parallel analytical procedure be adopted so that the results of the different analysis methods can be compared. The second method is often used as an enhancement to the first when the procedure adopted in the first method cannot justify the structure on its own.

We will present cases and critiques for all three of these procedures, as applied to a historic structure. In each case, we will review the conditions that make an analytical solution difficult along with the other alternatives for analysing this structure. We will then explain possible applications of each of the three general methods described above. Throughout, we will invoke the principle that structures built according to the building code prevailing at the time of construction are code-conforming. However, it is often necessary to review contemporary building codes when the structure is being altered or is showing distress.

7.2. Justifying the use of unorthodox analysis methods

Friedman (2008) describes typical fireproof floor construction methods dating from the late 1800s to the mid-1900s. These floor designs include iron or steel I-girders supporting arched tile floors, or tile floors in a flat arch configuration. There is a great variety of this type of structure in patented systems of one kind or the other. The infill may be 'stone concrete', cinder concrete (for lighter weight) or tile arches. By this account the tile systems may have an advantage of being favoured by building code officials at the time that these systems were widespread. The determination of the capacities of many of these systems was by testing and re-use of test results, rather than the development of a specific analytical procedure. Evidently, this reflects a bias towards the determination of load capacities empirically. The frequent experience of this author in determining the capacity of these floors indicates that the iron or steel beams alone are insufficient to justify the maximum loading on the floor. Although the fireproof infill was not intended as composite with the girders, the effect of these materials is to add strength to the floor system. The means of characterising this strength are as varied as the floor systems themselves. Access to testing results from the original construction of the floor may help in justifying the strength of the floor. A numerical structural analysis taking account of the apparent composite action of the floor and the girders has been used on occasion. On the other hand, Friedman (2008) considers these floor systems to be somewhat fragile and vulnerable to unthinking modifications. Clearly, any review of a floor system like this invokes the experience of the original builders and the experience of the preservation engineer reviewing this construction. This approach to the analysis of these floor systems has a dual character: it is directly empirical in using the evidence of the senses and of experience in recognising a floor

Figure 7.1 Segmental arch fireproof floor (From Kidder, 1916)

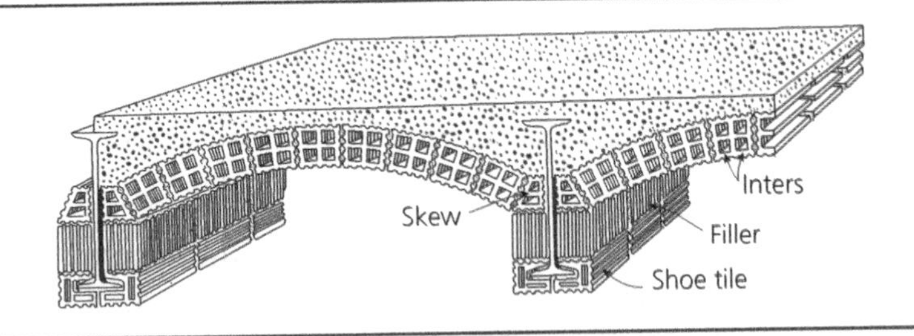

Figure 7.2 Flat arch fireproof floor (From Kidder, 1916)

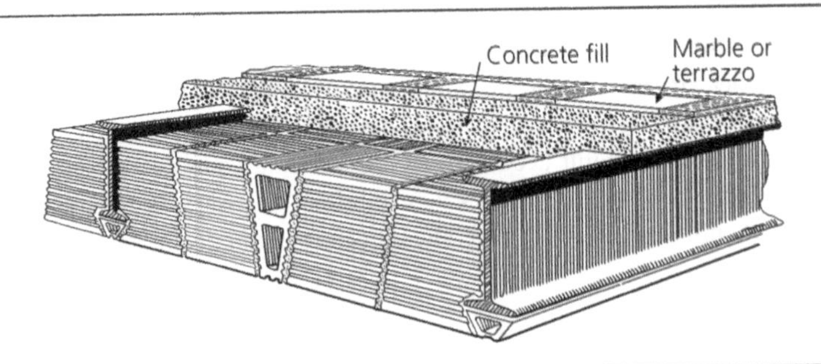

system that has significant strength, but in this method analytical methods are also used that may go beyond conventional analysis in determining the capacity of the floor, namely invoking composite action between the beams and the floor infill.

Friedman and Abdelfatah (2019) point out further difficulties in dealing with obsolete structural systems, such as fireproof floors, draped mesh concrete floors or masonry non-bearing exterior walls. There are no universal present-day proved analytical methods for accounting for the additional strength provided for these construction types. This can pose a dilemma when a structure must be assessed for new uses or for repairs. Two opposite approaches would be to accept empirical evidence that the structure has endured for a long time or to conduct an elaborate structural analysis to verify the strength of the structure. A frequently used, usually erroneous, option is to strengthen the structure without further investigation, because of a lack of faith in the first option and the cost of the second. Friedman and Abdelfatah (2019) outline some important middle ground of a reasoned evaluation of the structure based on the literature available on the design and construction of the original system and a reasonable explanation of the strength of the structure.

7.3. Development of non-standard analysis methods for historic structures

In the assessment of masonry arch bridges, engineers often attempt to find new ways to analyse a historic structure, rather than applying any modern procedures to such an assessment. The strength of masonry made of brick and stone is often undervalued in the assessment of nineteenth century structures. An article by Boothby (2020) on the application of the empirical design of arch bridges relates a series of successful, enduring long-span bridges to their proportions given by empirical rules.

In the US, there is little practical guidance on the analysis and design of masonry arches. There are a few unreasonably low allowable stresses available in the *Manual for Bridge Evaluation* (AASHTO, 2011), but no means of assessing the stresses in the bridge or the resistance available for the bridge. A review of the bridges in the US national bridge inventory (FHWA, 2022) shows that a large percentage of these bridges are evaluated by 'engineering judgement', which, as described in Chapter 6, is essentially an empirical method: saying that the bridge is subjected to a load magnitude without damage and judging that the bridge is capable of carrying that load.

In the UK, on the other hand, there is official guidance on the assessment of masonry bridges. A 1930s method for the analysis of masonry arches (known as the MEXE method) is embedded in the standards for the assessment of masonry bridges (Highways England, 2019). This method is approximate, conservative and contains subjective empirical reduction factors for the strength of a masonry bridge. Critiques of this method have been published (e.g. Gibbons and Fanning, 2012) but the method persists. The method is fundamentally empirical, employing empirical reduction factors for bridge strength, but the strength values that it begins with are not necessarily realistic. A publication appearing more recently is the CIRIA C800 guidelines (Gilbert *et al.*, 2022), which use partial safety factors for structural actions and resistance of the bridge as a basis for the assessment of masonry bridges. One of the case studies from this report is presented in Figure 7.3. This document, as is common for masonry

Figure 7.3 A masonry arch bridge assessment from Gilbert *et al.* (2022): (a) image of the three-span rail bridge; (b) details of the computer analysis assessment exercise

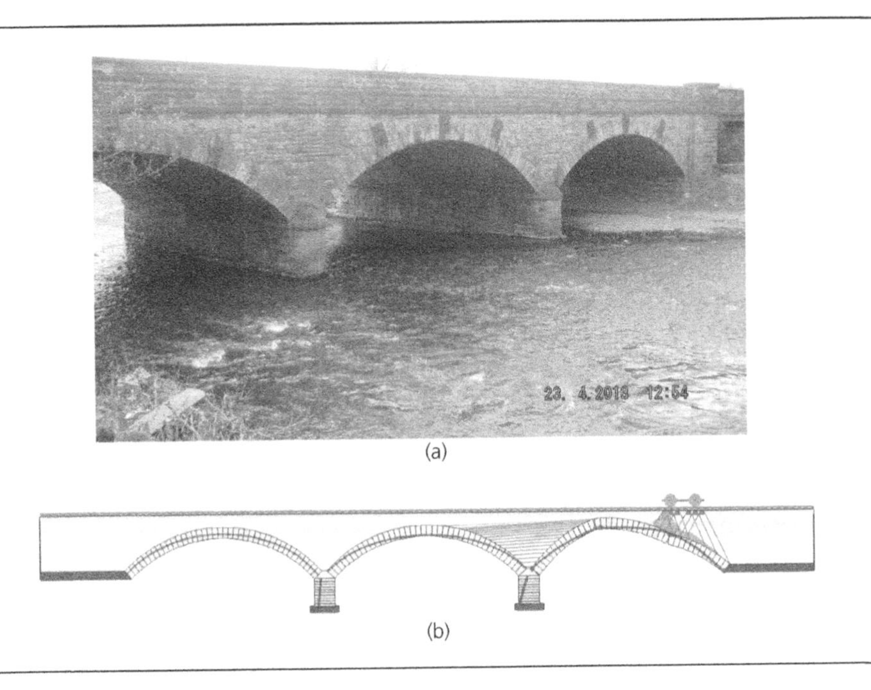

bridge assessment, is much more prescriptive in evaluating the span capacity of the bridge and less specific in how to evaluate the transverse effect. In Italy (CNR, 2015), a nationally disseminated document describes the determination of various factors of confidence for a masonry bridge (cracks, material properties, geometry and material deterioration), based on the condition of the bridge in general. These factors are fundamentally empirical. An observation method (also empirical) is described for the estimation of transverse effects.

7.4. Estimating the capacity of a structure based on longevity

The previously cited empirical procedures from the 1800s are also available and appear to be effective in predicting the required size of a masonry bridge. The results of the formulas of Rankine (1865) and Trautwine (1874) can be effectively compared with the dimensions of masonry and concrete arch bridges: both monumental bridges and much smaller structures. These methods give proportions – ring thickness to span and radius – that compare with intact bridges. What they lack for modern engineers is the ability to estimate loads – the formulas are just used to determine the correct proportions of a bridge, whereas the contemporary management of bridges demands a numerical assessment of the strength of the bridge. However, the load-carrying capability of the arch is seldom at issue. There are greater problems with the resistance of the spandrel walls. The most commonly observed failures of masonry arch bridges are movement of the spandrel walls, separation of the walls or horizontal cracking in the arch barrel below the inside of the spandrel walls. The estimate of the residual capacity under these circumstances is a matter of experience, particularly in understanding the experience of the structure and whether it is currently undergoing large loads. The maintenance of

Figure 7.4 Failure of masonry arch bridge, Blaine, OH, USA, due to transverse effects (Photograph by the author)

the residual capacity in this case also depends on the maintenance of the structure, particularly the redirection of stormwater from the bridge deck and away from the inside of the spandrel walls and parapet.

A cliché present in descriptions of the bridge assessment (Darden and Scott, 2005) is that the bridges were designed for lower loads than the present codes call for. The loads used in bridge design in the nineteenth century are of similar magnitude to the loads used at the present time. The largest load used in the design of nineteenth century bridges was an allowance for a crowd

Figure 7.5 Proof test of St Mary Street Bridge, San Antonio, TX, USA, c. 1890 (Image courtesy of Burndy Library, Darnell Collection, Huntington Library, San Marino, CA, USA)

of people on foot, reckoned as 4000 to 5000 N/m^2. This produces a lane load of 12–15 kN/m, which is larger than the lane loads used presently in the American Association of State Highway and Transportation Officials (AASHTO, 2011) specification for the design of highway bridges. Even the vehicles used for the design of urban bridges included a road roller, which was a very heavy piece of equipment, weighing up to 10 000 kg. This can be counted

Figure 7.6 Common failures of masonry arch in building walls: (a) failure due to spreading of abutments on thin arch with high span-rise ratio. The opening was filled in after the arch failure; (b) asymmetric failure of two brick arches, probably due to settlement of the central support

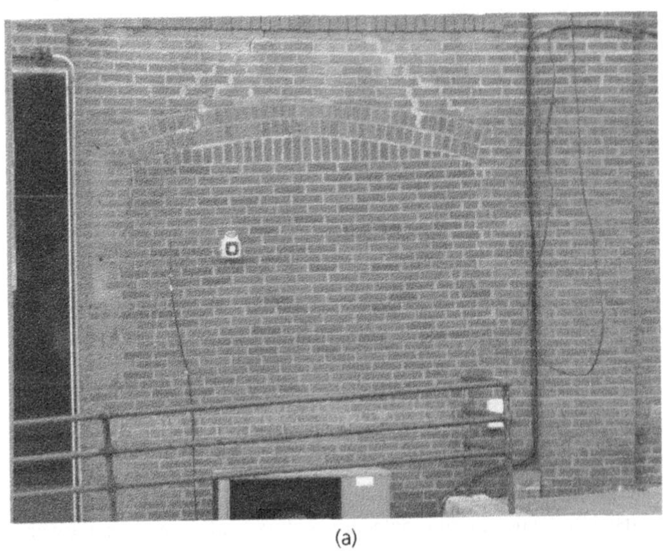

(a)

(b)

into the experience of load history of any bridge from the nineteenth century, especially bridges in towns and cities, where road rollers were in frequent use.

On the other hand, we often see masonry and concrete arch bridges determined as 'impossible to rate' because there do not appear to be standardised procedures for the analysis of these structures. This speaks either to the need to develop or to import procedures for the analysis of arch bridges or to become more skillful in the application of empirical analysis.

The usual procedure for arches in building façades is to accept them as sufficient, unless there is clear visual evidence of insufficiency. It would be very unusual to undertake an analysis of a brick arch that is not showing signs of damage. However, in this case the loads are constant and more predictable than for arch bridges. The largest problems in masonry building arches result from settlement of the supports or horizontal yielding of the supports. The former can happen nearly anywhere in the façade, while the latter tends to occur near the edges of the building. In this case the empirical aspect of the analysis and design is the ability to recognise damage and to determine which of these factors may be the cause.

Sometimes, empirical procedures can only be used in conjunction with analytical procedures. This applies to the most challenging areas for a preservation engineer, the assessment and development of lateral force-resisting systems for historic buildings. At least part of this exercise involves a visual and dimensional analysis of the extent of the bracing system. The lateral forces considered in building design include wind loads and earthquake loads. Wind loads are somewhat more predictable in their effects compared with the seismic loads described below, but can produce general pressures of up to 1500 N/m^2 on the frame of a building and localised pressures on the façades of a building of up to 2000–2500 N/m^2.

7.5. Wind loading

The assessment of the capacity of a building frame under wind loading involves tracing the load path through the walls, through the floor diaphragm and into the bracing system, which consists of rigid frames, braced frames or shear walls. The assessment of the load path is an exercise in empirical design. Some experience is necessary to have the skill to recognise which are the important parts of the load path. The recognition of the type of system which is suitable for differing heights of buildings is also an exercise in empirical design. Table 11.1 of *Empirical Structural Design for Architects, Engineers and Builders* (Boothby, 2018) gives an approximate distribution of the type of lateral force-resisting system that may be used for buildings of various heights and materials.

The book referenced above provides, in Table 10.2, a method for the empirical determination of the required length of shear wall. In this method, it is necessary to find the horizontal dimension of the building in a direction perpendicular to the direction of the wind (and to the direction in which the shear walls will be measured). The amount of shear wall needed per floor is expressed as a percentage of this horizontal dimension. Thus, for a 10-storey 50 m × 65 m building footprint, the total amount of shear wall provided in the short direction is 5% × 65 m × 10 storeys = 32.5 m and in the perpendicular direction 25 m. There is a similar suggestion that a single braced bay provides sufficient bracing for approximately five or six bays of similar size. While such empirical formulas are suitable for the initial determination of the adequacy of the shear walls in a building, it is barely suitable for a final assessment of a system.

On the other hand, the empirical assessment of area loads from wind effects on building façades is a matter of understanding in basic terms the anchorage of the elements under consideration and assessing this resistance in specific cases. As these elements are often hidden from view, it is necessary to review plans from the construction of the building, the loading history of the elements as determined from weather records, or to sample the anchorages and to attempt to determine the extent of the anchorage resistance. For brick façades, this involves brick ties or anchors. For other materials, such as aluminium panels, other methods of anchorage have been developed.

7.6. Seismic loading

Seismic loads present a significantly greater challenge for historic buildings due to the presence of unreinforced masonry and because of the general inability to make accurate predictions of the capacity of any of the elements, structural or non-structural, of a historic building. The general recommendations on the length of shear walls given above prevail only for areas with moderate seismic loads (seismic use category D or E). Where higher earthquake loads exist, empirical assessment encompasses a review of the structural and non-structural elements of the building and an estimate of which of these elements require more detailed engineering review. The understanding of seismic loading on historic buildings is a significant obstacle to their analysis and upgrading. The general empirical principles that would be observed in higher categories of seismic resistance are as follows.

- The avoidance of unreinforced masonry (or strengthening of unreinforced masonry)

  The strengthening of such systems often involves building a companion frame to support the wall.

- The avoidance of soft storeys

  Ground floors that have garage or shop front openings below more solid walls for apartments above are a type of structure that has a history of poor performance in earthquakes.

- The maintenance of a clear load path to the ground

  The load path includes horizontal diaphragms, vertical shear walls and braced frames.

- Solid anchorage between load-resisting elements

  Shear walls are bolted or cast monolithically with floor diaphragms. All ground floor lateral resistance elements are solidly attached to the foundations.

- Some form of ductility is built into the structure

  The yielding of steel, either concrete reinforcement or steel: if the element can be controlled suitably during failure it can contribute to the absorption of energy and reduce the severity of the earthquake effects on the structure.

7.7. Conclusions

Preservation engineering, the assessment of historic structures and their upgrading for adaptive reuse, new loading environment or general maintenance, relies to a large extent on empirical design and empiricist appeals. As many bridges and buildings up to 1900 were designed empirically, it is fitting to invoke empirical design in their later assessment. The longevity of a historic bridge or building may also be empirical evidence that it is designed properly. The engineer responsible for evaluating a historic structure considers that many materials and assemblies from the nineteenth century and earlier have strength that is not accounted for in modern analytical procedures – an example being composite action in fireproof floor systems – not part of the design intent, but useful in the capacity assessment. In the analysis of bridges, an understanding of the loading history is important. Until very recently, bridge assessment codes in the US and in the UK allowed the engineer to take the empirical evidence of satisfactory performance into account in bridge assessments. The exercise of empirical design in earthquake-resistant design reduces to a few simple principles: load paths, caution about unreinforced masonry, elimination of soft storeys and similar principles.

REFERENCES

AASHTO (American Association of State Highway and Transportation Officials) (2011) *Manual for Bridge Evaluation*. AASHTO, Washington, DC, USA.

Boothby T (2018) *Empirical Structural Design for Architects, Engineers and Builders*. ICE Publishing, London, UK.

Boothby T (2020) Empirical design of masonry arch bridges. *Journal of Architectural Engineering* **26(1)**: 6–14.

CNR (Consiglio Nazionale delle Ricerche) (2015) CNR-DT 213/2015: Istruzioni per la Valutazione della Sicurezza Strutturale di Ponti Stradali in Muratura. CNR, Rome, Italy, p. 108. (In Italian).

Darden C and Scott TJ (2005) Strengthening from within. *Public Roads* **68(5)**: 8.

FHWA (Federal Highway Administration) (2022) *Downloadable National Bridge Inventory*. https://www.fhwa.dot.gov/bridge/nbi/ascii.cfm (accessed 02/09/2022).

Friedman D (2008) Analysis of archaic fireproof floor systems. In *Structural Analysis of Historic Construction* (D'Ayala D and Fodde E (eds)). Taylor and Francis, London, UK.

Friedman D and Abdelfatah M (2019) Hidden strength in historic buildings. *Proceedings of the 2019 IABSE Conference*. New York, NY, USA.

Gibbons N and Fanning P (2012) Rationalising assessment approaches for masonry arch bridges. *Proceedings of the Institution of Civil Engineers. Bridge Engineering* **165(3)**: 169–184.

Gilbert M, Smith C, Cole G and Melbourne C (2022) Guidance on the assessment of masonry arch bridges – Part 1. CIRIA C800, London, UK.

Highways England (2019) Design Manual for Roads and Bridges: CS 454 – Assessment of highway bridges and structures. Highways England, London, UK.

Kidder FE (1916) *The Architects' and Builders' Pocket-book*, 16th edn. Wiley, New York, NY, USA.

Rankine WJM (1865) *A Manual of Civil Engineering*, 4th edn. C. Griffin, London, UK.

Trautwine JC (1874) *The Civil Engineer's Pocket-book*. Claxton, Remsen, and Haffelfinger, Philadelphia, PA, USA.

Boothby T
ISBN 978-0-7277-6633-5
https://doi.org/10.1680/edse.66335.081

Chapter 8
Case study II: Forensic engineering

8.1. Introduction

The purpose of forensic engineering is to identify possible, incipient or actual failures, to determine the likely cause of failure and to determine mitigation measures, either for the circumstances around a failure that has taken place or to avert a likely failure. The nature of the possible failures of a building may be structural, non-structural or a combination of the two. A structural failure might be the collapse of a cantilever balcony. A non-structural failure may be excessive humidity in a dwelling unit causing mould growth, while a combined failure might be a wall that was undermined by floodwaters and has collapsed as a result.

8.2. Forensic engineering

Forensic engineers often have significant involvement in mitigating possible failures. As an example, façade ordinances in many cities in the US require periodic inspection and maintenance of the façade to protect against falling pieces. Mohammadi (2021) lists the requirement for façade inspections in several major cities in the US. Typically, buildings over five storeys are required to be inspected under the supervision of a professional engineer at an interval of approximately five years. Chew (2021) analyses a large population of buildings in Singapore, a city which has recently implemented a façade inspection ordinance. The author presents a comprehensive view of possible building pathologies. He presents statistics from Singapore on the frequency of recurrence of certain faults in buildings, the most common of which is windows detaching from their frame and falling to the ground. His discussion of the previous failures is intuitive and is based on his experience and the experience of others. The main methods described are the implementation of checklists, review of commonly occurring failures and consulting with others on these topics.

In addition to reviewing specific failures or incipient failures, forensic engineering is also used to update existing codes, standards and practices by informing of previously unknown conditions that might pose a threat. Forensic engineers may lend their experience with building failures to the determination of building code provisions that may mitigate the risk of such failures.

8.3. Building science

Building science is the application of scientific principles, such as material science, heat and moisture transport and thermal science, to the design and assessment of buildings. Building science is often invoked by forensic engineers. Building science is a necessary part of the toolbox of a forensic engineer but, as we will show, building science is a mixed collection of appeals to various categories of science. Most of the work of a forensic engineer is based on experience-based review of conditions and experience-based recommendations for remedies.

A survey of building science is presented by Straube and Burnett (2005). The basic area of investigation is the effect of indoor and outdoor climate on building enclosures. The science of buildings encompasses solar angles, radiation and other effects, climate and temperature, imbalances in temperature and moisture between building exterior and interior, and the flow of heat and moisture that results from these imbalances. Building science is composed of enough fluid mechanics to understand the flow of air in and around buildings, enough heat transfer to understand the heat exchange between the interior and exterior of a building, and enough climate science to understand the physics of rain and wind. The topics explained by building science include the flow of heat and moisture through complex building envelopes, wetting and drying and freeze–thaw cycles of common building materials, control of condensation on the surface or within building materials, and the control of moisture through shedding or redirecting rainwater or storage of moisture within the building wall.

Using higher numbers of dimensions in this analysis results in more complex inputs and more complex results. The application of more complex models results in a greater number of input parameters being required and significantly greater difficulty in interpreting the results of the analysis. It is customary for building scientists in practice to rely on one-dimensional models unless a situation arises that clearly calls for the application of a more complex model.

The models on which building scientists rely describe the transport of heat, moisture and air through media such as the materials or surfaces used in building construction. They are developed to a greater or lesser degree depending on the complexity of the problem under investigation. A widely used computer programme for this building envelope analysis, WUFI, is most used in the one-dimensional version – that is, the assumption is that the building envelope consists of homogenous plane layers and that the moisture transport through the wall has the same characteristics in any location on the wall (WUFI, 2022). Any results from such a model are tempered with an understanding derived from experience about whether the results make sense and, very often, a search for sources of deviation of the modelling results from observed facts. Because models may not account for cracks or for gaps perpendicular to the direction of analysis, when the amount of observed moisture is greater than the predictions of a one-dimensional model, it usually results in a search for other sources of moisture: leaks admitting liquid water, cracks in mortar joints in a brick wall or other effects.

The methods employed by forensic engineers are described in several publications of case studies or class studies in forensic engineering. The study of organisation for façade inspections of Mohammadi (2021) describes the methods used by an engineer in conducting the review of a façade, including the tools needed for inspection, the organisation of the reports and other matters. The fundamental aspect of this programme is visual inspection, performed at arm's length from scaffolding whenever possible. Suspicions raised by noting breakage, missing pieces, cracks or discoloration in the inspection may trigger a recommendation for a testing programme. Other findings that may trigger a more extensive testing programme include probing sealant to check qualitatively for adhesion, probing mortar joints for firmness, using non-destructive techniques such as a thermal infrared (IR) camera to look for possible anomalies in a roof and so on.

An important part of a testing programme is the removal of existing pieces of the wall and inspection of the conditions of the anchorage. Laboratory tests may be ordered to determine strength, chemical changes, residual capacity, bond strength or other matters. These tests are secondary to the formation of an experience-based conclusion because of the observed conditions.

A review of the methodology of façade inspections allows us to describe more comprehensively the methodology of forensic engineering. There is usually a known or incipient failure to be reviewed. Most of the evidence is examined visually, with a hypothesis developed based on visual examination. The hypothesis is further tested by more careful visual scrutiny of the affected part, by the removal of material for laboratory testing, and by the assembly of this information into a conclusion about the probable cause. This cause is also tested for plausibility using the experience of the engineer conducting the investigation. Mitigation measures are developed based on the experience of the engineer. This description, in general, points to the application of empirical methods, occasionally informed by building science investigations, rather than an effort grounded in science and reason.

8.4. Forensic engineering methodology

Forensic engineering itself has two sides: the determination of the causes of a failure or the search for incipient failure. The second is significantly more difficult, given the unpredictability of failure and the difficulty in isolating conditions that cause failure in advance of the event. When research on possible building failure is focused on a single type of failure, say, detaching of terracotta tiles on a building façade, the scope of the investigation may be narrowed and the search can be confined to sampling and investigating tiles on a given building façade. A general review of the health of a particular structure, however, involves more factors, many of them unknown. In this case, the engineer relies largely on their own previous experience and the experience of others as outlined in the literature. The need to consult literature on obsolete building practices when reviewing a nineteenth century building structure has been described in the previous chapter. An understanding of methods of design used, especially in the design and placement of anchorages for façade elements, may be particularly useful.

In investigating structural failures, the forensic engineer often discards some of the building and material code-based procedures for the determination of the capacity of a structure and reviews the actual load paths within a building structure, which may differ from the load paths assumed in design. Most building codes incorporate significant safety factors, so that merely exceeding the allowable stress in a material or having an ultimate load greater than the nominal resistance of a structure are usually not sufficient to precipitate a failure. A forensic engineer must deal more thoroughly with the experience of the structure and more carefully estimate the actual resistance of the material without considering the safety factors that are embedded in the design values. A recent investigation of failure of residential truss floors (Broers *et al.*, 2021) focused on the occupancy and activities of the occupants rather than on the capacity of the trusses that failed, as the occupancy produced a larger load than the design load and the activity (dancing or jumping in unison) exacerbated these loads.

8.5. Building science methodology

Much of building science concentrates on the pathways for water and moisture to enter the building. There is less science involved in this aspect of understanding building than in many other areas of building technology. Instruments such as moisture meters may be used, but the principal method consists of locating pathways for water in the existing building and determining ways to limit the ingress of water into the building, working the way outwards from the interior point where water was detected to attempt to locate the point of ingress. The ability to complete this task effectively is primarily a matter of experience: experience with how water travels within roof membranes, experience with the effects of blocked roof drains, and experience in recognising signs of water ingress even where water is not found directly in the building. ASTM E2128 (ASTM, 2020) standards attempt to formalise methods to identify leakage pathways, but it is still fundamentally the engineer's knowledge and experience that drives the investigation plan. In normal practice, the leakage pathway, once identified, is verified by testing (Figure 8.1).

The science itself that is the basis of forensic engineering is an ad hoc assembly of ideas from other fields that may be relevant to investigations of building enclosures. The science that is employed is a small portion of a large variety of branches of science. These various sciences are brought to bear on specific problems that may have come to light in a particular building or, increasingly commonly, as part of the design process for a new building. In a sense, the choice of which science to use, or how to apply it, is made based on the experience of the particular engineer conducting the investigation.

The forming of hypotheses, arguably the most important part of a forensic investigation, is an empirical activity. It generally relies on minimal data, primarily visual, to try to form an idea of the issues that are present in a particular situation. In addition to the visual observations present, the knowledge and experience of the investigator are critical. An experienced investigator knows where to look for cracks, how to identify the results of water ingress, where to look for

Figure 8.1 Standardised leakage test (Image courtesy of RDH Building Science, Inc.)

Figure 8.2 Diagnostic leakage test (Image courtesy of RDH Building Science, Inc.)

water ingress and whether observed phenomena, such as rusting, are more likely due to atmospheric effects or to exposure to liquid water. The investigator's experience also dictates which hypotheses of probable cause of failure are likely and which can be ruled out altogether.

Figure 8.2 shows a different diagnostic leakage test of a building façade in progress.

The selection of tests or specimens to collect is fundamentally empirical. The budget of any investigation limits the amount of data that can be collected, and it is necessary for the investigator to use their experience to find a realistic amount of data to be collected that may result in a successful investigation.

8.6. Conclusions

We have seen, in general, that building science or structural mechanics are important tools, but are not entirely representative of the work of a forensic engineer. Although science is useful for explaining the observations and hypotheses of the investigator, the applications to be used are chosen in accordance with the investigator's experience. In the case of the application of structural engineering to structural failures, the investigating engineer uses their experience to strip away the safety factors and the specialised building code provisions to determine what are the likely conditions that are the cause of failure. In the application of building science, the investigator decides which aspects are pertinent – for example, whether a one-dimensional flow model for moisture vapour has any value given the state of cracking of the wall under investigation.

The principal tool of forensic engineering remains the senses of an experienced investigator. Many attempts are underway to enhance this tool: the use of drones for improved access, the application of augmented reality to allow the investigator to review more than just the evidence of their eyes and the use of machine learning to supplement the experience of the investigating engineer. However, forensic engineering and building science remain the products of experience, as carried out by an experienced investigator.

REFERENCES

ASTM (2020) ASTM E2128: Standard guide for evaluating water leakage of building walls. ASTM International, West Conshohocken, PA, USA.

Broers M, Bender D, Woeste F and Phillips A (2021) Residential floor failures from dynamic occupant loading. *Journal of the Performance of Constructed Facilities* **35(4)**: 9.

Chew MYL (2021) Façade inspection for falling objects from tall buildings in Singapore. *International Journal of Building Pathology and Adaptation* **41(6)**: 162–183. https://doi.org/10.1108/IJBPA-10-2020-0087

Mohammadi J (2021) Building facade inspection process: management and administrative matters. *ASCE Practice Periodical on Design and Construction* **26(3)**: 6.

Straube J and Burnett E (2005) *Building Science for Building Enclosures*. Building Science Press, Westford, MA, USA.

WUFI (2022) *What is WUFI*? https://wufi.de/en/software/what-is-wufi/ (accessed 05/09/2022).

Boothby T
ISBN 978-0-7277-6633-5
https://doi.org/10.1680/edse.66335.087

Chapter 9
Case study III: Design of a wood-framed building

9.1. Empirical design of a timber-framed residential building

In this case study, we will examine a prototype three storey (one level at grade and two storeys above grade) residential building constructed with dimension lumber. We will first determine the sizes of the elements empirically, using the rules given in the author's previous work (Boothby, 2018), and compare selected elements with the results of analytical design according to the *National Design Specification* (NDS) (AWC, 2018) and according to Eurocode 5 (BSI, 2004). The International Building Code (ICC, 2021), section 2028, includes a set of prescriptive requirements for conventional wood-framed construction, which we will also examine.

The building is a block of flats with separate exterior entrances and stairs. Each block contains four two-bedroom units per storey, 12 units (per block) in all. The construction will be timber, either engineered wood products or dimension lumber. The roof framing will be plate-connected wood gable trusses. This configuration of apartments, shown in Figure 9.1, is very typical for this type of construction and occupancy. An example of a block of flats that uses this layout is shown in Figure 9.2.

The empirical design of the floors is as follows. There are three possible schemes for supporting the floor: clear-spanning along the 6500 mm length of the dwelling unit (parallel to the access stairs), clear-spanning the 7000 mm width or spanning in the 7000 mm direction and providing an interior bearing wall dividing the sleeping rooms from the remainder of the apartment. For this two-span case, the spans are 3000 mm and 4000 mm. The entry porch has a width of 2500 mm to allow sufficient space for a stair. If the joists are to span the full width of a dwelling unit, then it is necessary to use engineered wood products of some kind: wood I-joists, wood truss joists or 4 × 2 wood trusses. The last alternative is probably ruled out by the small scale of this project and the relatively short span: 4 × 2 trusses (trusses assembled with top chords of 2 × 4 nominal (38 mm × 75 mm) lumber laid flat) are more suitable for spans from 7 to 10 m. Clear-spanning with wood I-joists, based on US span tables (Boise-Cascade, 2023), will require a depth of 360 mm and a spacing of 490 mm. The UK interactive span tables (James Jones, 2022) recommend an I-joist product with a depth of 400 mm and a spacing of 480 mm. Dividing the span allows the use of dimension lumber, with a likely depth, for a span:depth ratio of 15:1, of 275 mm (2 × 12 US). In both cases, the exterior entry porch span is framed with 38 mm × 200 mm dimension lumber. The direct application of an empirical span/depth rule does not identify a spacing for the structural joists. However, the use of a span table can allow the determination of an approximate required spacing.

Figure 9.1 Floor plan of four dwelling unit cluster for example wood-framed building. The building in the example has two storeys above the ground floor (Illustration by Lisa Han)

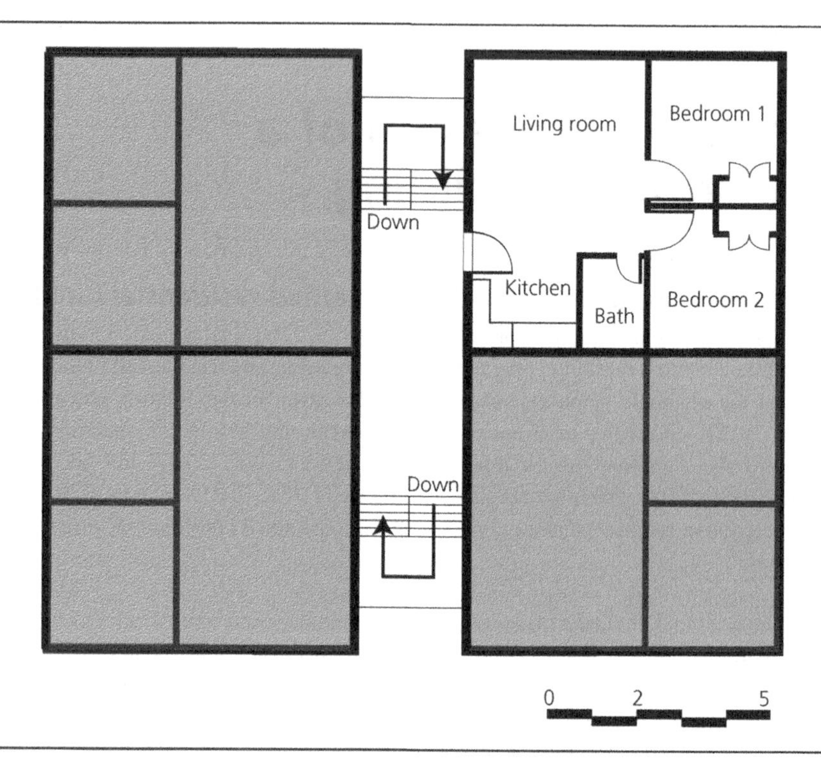

Figure 9.2 An example of a block of flats built according to the model of Figure 9.1 (Photograph by the author)

Roof framing consists of plate-connected wood trusses assembled from dimension lumber. The roof trusses are simple gable trusses with a roof pitch of 3:12 (approximately 14°). These trusses will clear span across the length of two flats, approximately 13 000 mm. The specific design of plate-connected timber trusses is delegated to the truss manufacturer. Typically, trusses of this span normally are in a four-panel fink truss configuration, have top and bottom chords approximately 38 mm × 150 mm and web members 38 mm × 100 mm. The trusses extend 600 mm beyond the wall line to form an eave overhang.

Most of the walls are bearing walls assembled from 38 mm × 100 mm studs. Using the proposed 6500 mm clear span scheme, an interior bearing wall separates two dwelling units. This wall has the greatest axial load; however, it has practically no out-of-plane lateral load. The exterior bearing walls are exposed to wind loading but resist approximately half as much axial force as the interior bearing walls. The interior bearing walls can have a stud spacing of 400 mm in the upper two floors, but the stud spacing will most likely have to be reduced to 300 mm at the ground floor or the studs doubled (see Boothby, 2018). The studs in exterior bearing walls may need to be larger, say 38 mm × 125 mm. The wind loads normal to the wall induce bending in the exterior wall studs and the larger size may be necessary to resist this bending. Moreover, energy codes usually require wall insulation thicker than the width of the smaller studs. These walls, especially if the studs are the larger size, will have single studs spaced 400 mm apart.

An empirical rule gives the spacing of the interior bearing walls studs as doubled 38 mm × 100 mm studs at a spacing of 300 mm to support three floors (or two floors and a roof, where necessary) or single 38 mm × 150 mm studs at a spacing of 300 mm for an exterior wall.

Empirical design of the foundation is based on widespread practice for low-rise residential buildings. The foundation will consist of strip concrete footings, approximately 300 mm × 600 mm in cross-section, supporting a 200 mm concrete masonry unit stemwall, which extends to 150 mm above grade. Spot loads from posts supporting the stairs may have a 600 mm square concrete footing. The ratios of the length of shear walls to the width of the perpendicular building face exposed to lateral forces given in Boothby (2018), along with standard practice, dictate that most of the walls of the building will act as shear walls for lateral force resistance and that this will be sufficient in most cases.

9.2. Empirical design summary

Floor framing: wood I-joists clear-spanning 6500 mm, depth 400 mm, spacing 480 mm

Alternative floor framing: wood dimension lumber 38 mm × 275 mm with interior bearing wall in each dwelling unit (spans 4000 mm and 3000 mm)

Bearing wall framing: 38 mm × 125 mm exterior walls, 38 mm × 100 mm interior walls

Wall stud spacing: 300 mm at ground floor, 400 mm above

Roof: plate-connected timber gable trusses, spanning 13 000 mm

Entry porch: 38 mm × 200 mm timber joists

Foundation: 600-mm-wide strip footings under bearing walls.

9.3. Analytical design of the case study wood flats

A similar design for this structure can be completed analytically in conformance with the provisions of Eurocode 5 (BSI, 2004) for the UK or following the NDS for a similar design based on US standards (AWC, 2018). Much of the information needed to complete a routine design is available in span tables (such as span tables for dimension lumber and engineered wood product joists) or is not completed by the engineer of record (such as the detail design of roof trusses for timber framed buildings). Reference will be made to available span tables.

9.3.1 Floors: dimension lumber

The maximum span is 3700 mm. According to the span tables in the WWPA (2008) *Product Use Manual*, Table 7, 38 × 250 spruce-pine-fir no. 2 spacing 450 mm is sufficient for this span in a residence. To use UK span tables, for a superimposed load of 1.5 kN/m^2, grade 16 timber, a 3700 mm span is achievable using 45 mm × 220 mm lumber, spacing 450 mm (Rightsurvey Co., 2018).

9.3.2 Floors: engineered wood product

The selection process is the same as it was for empirical design. Review a manufacturer's table to determine appropriate sizes and spans: 16-inch (400 mm) BCI 6500S 1.8 at 19.2-inch (490 mm) spacing. Floor is 0.75 inches (18 mm) OSB. An online calculator (Timber Beam Calculator, 2023) gives a similar solution of WS 400 Easi-joists 400 mm depth, with 72 mm × 47 mm chord, spacing 600 mm for an engineered wood product with metal trussed webs.

9.3.3 Interior bearing wall

Using the engineered wood products selection above, the interior bearing wall supports the equivalent of the width of a single unit. Design according to the NDS will use the allowable stress design (ASD) procedure, so the loads for this example will be left unfactored. The total load per floor on the bearing wall is 2500 N/m^2 × 6700 mm = 16.5 kN/m/floor. To support all three floors a total resistance of the interior bearing wall of 49.5, say 50 kN/m, is required. The calculations given below for NDS and Eurocode 5 verify the strength of a single stud to resist this force.

9.3.3.1 National Design Specification, (AWC, 2018)

The maximum axial compressive capacity of a single stud is found from Equation 3.7.1 of the NDS, using the ASD approach. The total load on the wall at the lowest level is 50 kN/m. We will try to design the wall using spruce-pine-fir no. 2 studs with a spacing of 300 mm (F_c parallel to grain 7 MPa; E_{min} = 2.76 GPa). It is first necessary to calculate the critical buckling stress in a single stud as

$$0.822\,E_{min}/(L_e/d)^2 = 3.01\,\text{MPa} \qquad (9.1)$$

so, the buckling strength of a 38 mm × 90 mm stud is 10.3 kN. The column stability factor is calculated as $C_p = 0.387$, giving a maximum allowable stress of 2.71 MPa, so a maximum load of 9.26 kN. Thus, a minimum of 5.4 studs would be required per meter. In practical terms, this results in doubled studs at a spacing of 300 mm, similar to the empirical design above.

9.3.3.2 Eurocode 5 (BSI, 2004)

To apply Eurocode 5, it is necessary to multiply the dead and live loading above by partial safety factors: 1.35 for dead load and 1.50 for live load. The distribution of loads of 2000 N/m^2 live and 500 N/m^2 dead results in a load on the interior bearing wall of 24.2 kN/m, so the load on the bearing wall at the lowest level is 72.6 kN/m. In calculations based on Eurocode 5, using wood studs of class 16, the tabular compressive stress, in the absence of instability, is 17 MPa, while the fifth percentile modulus of elasticity is 5400 MPa. Modified by k_{mod} and a partial safety factor γ_M gives a usable strength of 10.4 MPa. The slenderness ratio for a 38 mm × 100 mm stud, braced by the wall covering in the weak direction, is 96. The Euler buckling stress is therefore 5.78 MPa. λ_{rel} = 1.71. Based on Figure 6.5 in Porteous (2013), $k_c \approx 0.4$, so that finally the strength of the stud in compression is 0.4 (3420 mm^2) (10.4 N/mm^2) = 14.2 kN. For a bearing wall supporting two floors and a roof, the approximate factored wall load of 24.2 kN/m results in a load of 72.6 kN/m at the lowest level. Then the wall requires approximately 5.11 studs per meter, resulting in the same design as above, doubled 38 mm × 100 mm studs spaced at 300 mm.

9.3.4 Roof structure

The usual procedure for roof trusses is for the engineer of record to determine the shape of the trusses and to identify the loads that the trusses are to be designed for. The truss manufacturer is customarily tasked with determining the exact configuration of the truss and with choosing the size of the members and the truss plates. In this case, the decision is the same as in the empirical design: gable trusses with a top chord slope of 3:12. The loads to be specified are superimposed dead load, probably 0.75 kN/m^2, and snow load, depending on the geographical location.

9.3.5 Comparison of the three methods

The resulting design is practically the same based on the three methods considered. The only potential variation is in the spacing of the structural elements. The joist spacing determined by empirical and analytical design is the same (450 mm) for both designs, while the spacing of the wall studs is left open in the empirical design and is 300 mm spacing (doubled 38 × 90) in the interior wall and 400 mm spacing (single 38 × 150) for the exterior walls. While analytical design allows the determination of both the size and spacing of the structural elements, it is reasonable to assert that the resulting designs are similar. When the design of certain structures, such as multifamily housing, becomes routine the designs can be established by empirical design, with analytical design used as a tool for verification of the design.

REFERENCES

AWC (American Wood Council) (2018) *National Design Specification.* AWC, Leesburg, VA, USA.

Boise-Cascade (2023) *I-Joist Span Tables. Eastern Commercial Guide.* https://www.bc.com/span-and-size-charts-for-bci-joists/ (accessed 05/06/2023).

Boothby T (2018) *Empirical Structural Design for Architects, Engineers and Builders.* ICE Publishing, London, UK.

BSI (2004b) BS EN 1995-1-1:2004: Eurocode 5: Design of timber structures – Part 1–1: General – Common rules and rules for buildings. BSI, London, UK.

ICC (International Code Council) (2021) International Building Code 2021. ICC, Washington, DC, USA.

James Jones (2022) *JJI-Joists.* https://www.jamesjones.co.uk/products-and-services/engineered-timber/jji-joints/ (accessed 05/06/2023).

Porteous J (2013) *Designer's Guide to Eurocode 5*. ICE Publishing, London, UK.

Porteous J (2013) *Designer's Guide to Eurocode 5: Design of Timber Buildings EN 95-1-1*. ICE Publishing, London, UK.

Rightsurvey Co. (2018) *Easi Span Tables*. https://www.rightsurvey.co.uk/wp-content/uploads/2018/07/Easi_Span-tables.pdf (accessed 30/05/2023).

Timber Beam Calculator (2023). https://www.timberbeamcalculator.co.uk/en-gb/span-tables/ (accessed 23/03/2023).

WWPA (Western Wood Products Association) (2008) *Product Use Manual*. WWPA, Lake Oswego, OR, USA.

Boothby T
ISBN 978-0-7277-6633-5
https://doi.org/10.1680/edse.66335.093

Chapter 10
Case study IV: Reinforced concrete foundation

10.1. Introduction: semi-empirical design

For general guidance on the size of elements of a foundation system, the reader may refer to *Empirical Structural Design for Architects, Engineers and Builders* (Boothby, 2018). Chapter 8 (Empirical design of foundations) in that book invokes semi-empirical design, which is design based on a simple calculation of load against size, with the rest of the features of a structure left to empirical rules or to the knowledge of the builder. In the case of foundation design, an estimate of the size of a spread footing can be made based on a gross estimate of the load on the column supported by the footing divided by an estimate of the allowable soil bearing pressure.

An ordinary commercial building having 8.5 m × 8.5 m bays and a floor load of approximately 4000 N/m^2 has a column load of approximately 300 kN per floor. For a building with three floors above grade, the load on a spot footing will be approximately 1200 kN. The minimum allowable bearing pressure for a competent soil is 100 kN/m^2. The subgrade of many building foundations supports twice this value. So, given a four-storey office building of the bay dimensions above, using a minimum allowable bearing pressure value, the required footing size is 12.0 m^2, or a footing size of 3500 mm square. This is a reasonable reference value for the size of the column footings in a commercial building. Smaller footings are often possible, especially if the allowable soil bearing pressure is greater.

In the case of a wall footing supporting a 200 mm concrete masonry bearing wall, the minimum strip footing width is 600 mm. As a result, the capacity of the minimum size footing is 60 kN/m, given the minimum allowable soil bearing pressure above. For a block of flats with precast hollow core plank floors, the load on an interior bearing wall is approximately 20 kN/m per storey. It is apparent that the minimum size wall footing can support at least three floors of concrete masonry unit (CMU) wall and precast hollow core plank floor. In the timber construction of the previous chapter, this size of wall footing can support additional floors.

10.2. Empirical design of a shallow foundation system

For a building with a shallow foundation supported by a bearing stratum of soil that will support a minimum allowable bearing pressure, the design of the foundation elements follows the discussion above very closely. The size of spread footings, which are in proportion to column loads, can be estimated to be approximately 3.0 m^2 per floor supported – that is, four floors (or three floors and a roof) supported by a spread footing requires a total footing area of 12 m^2 (3500 mm × 3500 mm), as described above. Any strip footings required for interior or

Figure 10.1 Placement of concrete in a typical spread footing foundation (Photograph from Energy.gov public domain at Wikimedia Commons, https://commons.wikimedia.org/wiki/File:DX_Groundwater_Treatment_Facility_concrete_pour_(7582956780).jpg)

exterior bearing or non-bearing walls can be simply built to the minimum size: the width of the CMU or concrete wall that the footing is supported plus 200 mm on each side. For 200 mm CMU, the required footing width is 200 mm projection plus 200 mm stemwall plus 200 mm projection. Except in special circumstances, the slab on grade is 4 inches of 5-inch thick concrete bearing directly on the subgrade for the building.

The reinforcement in a spread footing is designed on the basis of a cantilever beam, with the reinforcement placed in the bottom of the footing. Due to the limited size of spread footings, the amount of reinforcing required is usually the minimum, 0.2% of gross concrete area in the US. In Eurocode 2 (BSI, 2004), the minimum reinforcement is different for different circumstances, but generally less than 0.2%. For a 375 mm thick footing, 0.2% reinforcement results in a 20 mm dia. bar spaced by 300 mm. The flexural capacity of this minimum reinforcement is often adequate for the size of the footing. The bending moments for a 1525 mm projecting footing 1/2 (3500 mm footing dimension – 450 mm pier) and a 100 kN/m^2 bearing pressure are 116 kN m/m of width, or 407 kN across the entire 3500 mm width. The capacity of the minimum footing reinforcement in a 375-mm-thick footing is approximately 124 kN m/m. Higher bearing pressures require a shorter cantilever for the design of the spread footing. The thickness of the footing may be able to be reduced accordingly.

10.3. Sizes of elements in the foundation system

Based on the above semi-empirical design of a conventional spread footing system, we can state some general proportions for the design of such a footing in general. The spread footing thickness is recommended to be 1/12 its greater dimension. For instance, the 3500 mm square footing designed above would have a thickness of 300 mm. This is consistent with the recommendations (Boothby, 2018) for a concrete cantilever beam to have a span:depth ratio of

6:1, as the footing can be visualised as a double cantilever. Reinforced concrete piers above the footing require a minimum size of 450 mm × 450 mm. Reinforced concrete foundation walls require a 300 mm thickness. These walls can also be built of grouted concrete masonry with a thickness of 200 or 300 mm. Slabs-on-grade in commercial and residential construction are 200–250 mm thick. Basement walls that are in a retaining condition require support at the top and bottom and a height thickness ratio of approximately 12:1. Foundations for commercial buildings often have reinforced concrete grade beams that span from footing to footing and support wall loads or other loads from above. Their depth is determined both by flexural capacity and the predicted maximum frost depth. The frost depth, of course, varies depending on local conditions. The design for flexural capacity requires the same span:depth ratios as for superstructure concrete beams, approximately 20:1. In most temperate climates, the frost depth requirement dictates the actual required depth of this element.

10.4. Deep foundation system

Deep foundations are often required based on the local soil conditions. Deep foundations consist of drilled piers, auger cast piles, driven piles or micropiles. The length and size of these elements is dictated by soil capacity as interpreted by the project engineers and their consultants. It is not possible to advance empirical rules to cover these highly variable cases. However, all types of deep foundations have piles or piers individually or in groups or clusters located beneath the interior and exterior columns. A reinforced concrete pier or pile cap is placed over these elements and supports grade beams, sized by the principles given above. The size of the pile caps depends on being able to cover the irregularities of the pile cut-offs and to be able to be cast efficiently. A minimum of 600 mm length or width prevails and an overall size:depth ratio of 4:1 to 6:1 prevails. A triangular group of three piles with a clearance of 1200 mm between piles would require a pile cap of 2000 mm minimum lateral dimension; thus, a pile cap thickness of 500 mm to 750 mm would be required.

10.5. Discussion of foundation system design

Conventional building foundation design can be accomplished with reasonable completeness and reasonable success by simple empirical means. This is not meant to suggest that empirical design should be the sole means of completing the design of foundations in a contemporary building. However, the sizes determined by empirical design are valid estimates of the design requirements for most conventional buildings. In the case of buildings of small scale, such as one to three-storey residential or commercial buildings, there is really very little analytical design that needs to be done in completing a foundation design. Strip footings 600 mm wide under bearing walls or 1000 mm square spread footings under columns are suitable for most of the loads that one would expect to encounter in a building on this scale. Most of the other features of a building foundation – stemwalls, grade beams piers and other features – are simply built conventionally without required analytical input.

REFERENCES

Boothby T (2018) *Empirical Structural Design for Architects, Engineers and Builders*. ICE Publishing, London, UK.

BSI (2004) BS EN 1992-1-1:2004: Eurocode 2: Design of concrete structures – Part 1–1: General rules and rules for buildings. BSI, London, UK.

Boothby T
ISBN 978-0-7277-6633-5
https://doi.org/10.1680/edse.66335.097

Chapter 11
Continuing uses of empirical design

11.1. Introduction

The future of structural engineering promises continuing value for empirical design. One of the notable features of empirical design, as used during transitional periods between empirical and rational design (such as the late 1800s), is the combined use of both empirical and rational methods: empirical design to keep the designer in control of the design process and rational design to determine suitable sizes for the given components. This is a balance, often lacking in present day design, which can be restored by a better understanding of the relationship between analytical and empirical design.

Most of the design work including, in some cases, the determination of the configuration of the structure is turned over to analytical methods, which increasingly make use of computer programmes. Many of the computer programmes used for structural design embody rational/analytical thinking to the exclusion of the incorporation of simple experience by a designer. However, the growth area in the applications of computers directly to structural design is more likely to embody the concepts of machine learning, described below (Málaga-Chuquitaype, 2022).

The application of machine learning, or artificial intelligence, to computer programmes has potential for the incorporation of aspects of empirical design. Many of these applications simply examine outputs compared with inputs without using 'physics-based' modelling so that a programme of this type could be characterised as computerised empirical design. These programmes also make use of the resources of a skilled operator or designer to invest ideas of common practice, experience with a particular building system, experience in the analysis of a building system or other empirical values.

11.2. Balanced empirical and rational design

The author has previously identified the nineteenth century as a period in which engineers maintained an effective balance between analytical and empirical design (Boothby, 2015). The textbooks of this time period show that the authors were interested in modern arguments but include questions more relevant to the practice at that time. George Fillmore Swain (1927), in writing the three-volume set of books on structural engineering, presents very advanced ideas in the computation of forces in trusses, in graphic statics, in theories of earth pressure, but he also describes masonry structures in very practical terms, using the terminology of the masonry construction industry, and presenting detailed specifications for the construction of bridge masonry. The encyclopaedic *A Treatise on Masonry Construction* of Ira Osborn Baker (1907) is equally divided between theory and practice. The design principles advanced often include empirical ideas: for instance, the proportional rules of Rankine (1865) and Trautwine (1874) are invoked for the design of arches. The nineteenth century textbooks describe construction as an integral part of the structural design process. Baker (1907) gives a long account of the various theories of computing the pressure and the line of pressure in an arch, and further gives an equally thorough account of the construction of arches. Standard designs and details of smaller railroad arches are given, and construction and falsework diagrams are given for the larger arches.

The result of the training and the practice described above is that the practitioners of engineering were equally comfortable with the design of a bridge or building and with supervising its construction. Design was not an abstract exercise in mathematics but was accomplished with a sense of the character of the finished artefact. The design of a masonry arch or a wrought iron truss began with a vision of the finished work: its shape and thickness in the case of the arch, and its form and members in the case of the truss. The analysis was conducted on a complete conception of the final work and served to verify the empirical intuitions of the designer.

The restoration of design methods balanced between empirical and analytical requires the development of a basic understanding of empirical design as a legitimate design method, not simply as a procedure for checking the results of an analytical design. This requires that we revert in our thinking to recognise that engineering decisions can be made without an understanding of the laws of nature governing a process, or without even accepting the existence of a governing law of nature. We have shown in Chapter 9 that, to design a wood structure, we do not need to recognise the process of axial force, bending, buckling or moment magnification. It suffices to ensure that the height to width ratio of the column be sufficient. At the point that this understanding is reached, empirical design may become an equal partner to analytical design, much as it was in previous centuries.

11.3. Education

The role of empirical design in engineering education can be greatly expanded. Structural engineering education is sometimes criticised for being too detail-oriented or using assumptions that are unrealistically idealised. This critique of engineering education is reviewed and discussed, for instance, by Solnosky (2020). Instead of a study of the design of the pieces of structure-beams, columns, slabs – the focus of empirical design is the entire structure. While not necessarily furnishing an entire design, the merit of empirical design is that it permits the very rapid assembly of a plausible design. As an example, in the design of a three-storey wood-framed building in Chapter 9, the understanding of the system of the building preceded the sizing of the individual elements. From there, individual elements (floor deck, joists, girders, bearing walls, roof trusses) were configured and sized in relation to their role within the structure and their span. The design is based on global premises of structural performance, without interference from attempting to follow detailed design requirements for individual members. This difference in approach is exemplified by the design of the load-bearing stud walls. Empirical design recognises that a load-bearing wall is a system of 38 mm × 110 mm wood studs at a spacing of 300 to 600 mm. The detailed structural design requires the investigation of the buckling behaviour of the studs to assign a load-carrying capacity and to determine the exact spacing. For the exterior bearing walls, this requires analysis of axial force and bending for combinations of gravity and wind loads. Although the detailed design of the structure may be necessary, this kind of detailed thinking can cloud student understanding of the overall objective and overall expected results of the design. The results of the empirical design can be used as a means of checking the results of a more rational design, or a design that is more concerned with code compliance. Regarding the over-idealisation of building loads and assemblies used in teaching structural design, empirical design provides a means of looking beyond the complications of loading or of the assembly of elements and choosing a design that is dimensionally consistent. According to Hanson (2022), the training of structural engineers includes minimal strategies for assessing the reasonableness of the calculations performed or of the designs completed. Although this article is about structural analysis, a similar criticism could be levelled at design education. The careful weaving of empirical design components into steel, concrete, wood and masonry design education could reassert the place of empirical design.

11.4. Machine learning

One of the more modern approaches to the automation of data collection and structural design takes an empirical viewpoint. The use of 'data-based' as opposed to 'physics-based' modelling in the development of neural networks and machine learning algorithms is an exercise in empiricism. This method is often favoured as it requires only a large set of inputs and outputs to a process, and so does not require the development of elaborate and potentially error-prone models to make useful predictions of a process. In this procedure the inputs of a process are simply reviewed by a computer-based algorithm and compared with the outputs in a process called 'training' to find the probability-based rules to relate new inputs to projected outputs. An example might be training a computer-vision process to review multiple passages of trucks of different axle weights and configurations over a bridge girder in order to infer the weight and axle configuration of a truck for which these values are not known in advance. When this process is simply statistically based, the procedure is strictly empirical, in that there is no consideration of any theory of beam bending, of the suspension of the weight on the truck or of the response of a beam to dynamic loading. These theories can be quite complex on their own and extremely complicated when considered together. Instead, the exercise becomes a matter of comparing inputs (truck weight and wheel configuration) with outputs (girder deflected shapes) and using a large dataset of these factors to predict the outcome. Experience, whether the experience of a person or the experience of a computer, makes the judgement in this procedure.

The argument against using such statistically based models is that they have little additional scope beyond the problem that they were tasked with – the previous example, for instance, could not be used to investigate loads on a girder in a building. The argument for the use of this method has been made throughout this book – combinations of physics-based models are unnecessarily complicated, are often just as difficult to adapt to new circumstances, require the input of many incompletely known parameters and lose the clarity of a simple connection of one observation to another.

11.5. Re-introducing empirical design into contemporary practice

In the 1800s engineers had at their disposal both empiricist methods and developing rationalist methods. By the next century, the profession had embraced rationalist methods. However, the tools that contemporary engineers use for structural design also allow a choice between the application of empiricist or rationalist methods. In a way, the application of computers has put us at a similar crossroads: do we want to embrace rationalist methods to the exclusion of experience or do we want to strike an appropriate balance between the two modes of thinking? On many occasions, the success of engineering has been the ability to mediate between these two modes of thinking. At the point where an engineer embraces rationalism only, they risk losing contact with the structures that they design. On the other hand, a fully empiricist method of designing requires rethinking design methods for each emerging type of structure. Computer programmes, in some ways like the educational models described above, represent simplifications of the actual material and the actual assemblies. Even in the most sophisticated models, some elements are omitted, either for the sake of simplicity or on account of ignorance of the actual nature of the material. These omitted factors may have an important influence on the structure. As such, it is significantly better to understand how the structure is assembled and what are the actual materials of which it is composed. As in the example in Chapter 9, a computer model of wood studs, no matter how many properties of the wood are included in the model, would tell us very little of use in the design of the stud, in which the critical factors are the location of the connectors, the axial force capacity of the stud and the support conditions of the stud. What is needed, even when a stud in the bearing wall is designed rationally, is an overall understanding based on experience, of the likely strength of the stud, of its mode of attachment, of its likely out-of-straightness and other factors that are the result of experience, not analysis. As a structure grows in complexity beyond the simple wood framing exercise, the analytical information acquires greater value, but it is still necessary to bring in the experience of how the elements are connected, what is the likely capacity of the connections, in what parts of the structure problems are likely to occur. Whether or not the design is explicitly empirically based, it depends on empirical information for its success. The tools for training engineers in analytically based design are everywhere. The tools for training engineers in empirically based design and in an empirically based outlook need improvement to promote a complete outlook on the design of structures.

REFERENCES

Baker IO (1907) *A Treatise on Masonry Construction*, 9th edn. Wiley, New York, NY, USA.

Boothby T (2015) *Engineering Iron and Stone: Understanding Structural Analysis and Design Methods of the Late 19th Century*. ASCE Press, Reston, VA, USA.

Hanson J (2022) Teaching students how to evaluate the reasonableness of structural analysis results. *Journal of Civil Engineering Education* **148(1)**: 8.

Málaga-Chuquitaype C (2022) Machine learning in structural design: an opinionated review. *Frontiers in Built Environment* **8**: 815717. See https://doi.org/10.3389/fbuil.2022.815717 (accessed 31/05/2023).

Rankine WJM (1865) *A Manual of Civil Engineering*, 4th edn. C. Griffin, London, UK.

Solnosky R (2020) Mini project explorations to develop steel and concrete gravity system design skills. *ASEE Virtual Conference*, 22–26 June 2020. American Society for Engineering Education.

Swain GF (1927) *Structural Engineering*. McGraw-Hill, New York, NY, USA.

Trautwine JC (1874) *The Civil Engineer's Pocket-book*. Claxton, Remsen, and Haffelfinger, Philadelphia, PA, USA.

Boothby T
ISBN 978-0-7277-6633-5
https://doi.org/10.1680/edse.66335.101

Chapter 12
Conclusions

Empirical design was practised as a method for determining the configuration, material, span and size of structural systems and elements to at least the start of the nineteenth century. Any building from classical Greece or Rome, or other civilisations of the first millennium, has been designed by empirical methods. Although the ancient Greeks and Romans and the medieval architects had theories of mechanics available to them, these were the province of philosophers and had little to say to practical workers. In spite of the occasional recourse to a scientific understanding of dynamics – why heavy objects fall, for instance – ancient natural philosophers did not apply their understanding of natural phenomena to building construction. Even when the mechanics of the early nineteenth century became well developed to be used in evaluating stresses in building elements, this understanding did not develop into a clear, widely used procedure for design of a beam until mid-century. Until the end of the nineteenth century, empirical methods continued to be used for the design of many structures, including speciality metal, stone and wood structures that could be designed by the review of pattern drawings.

Although, in the contemporary environment, rational or analytical methods of engineering are much more widely practised than empirical design, empirical design has endured as an important adjunct to engineering design. The basis of most engineering principles is empiricist appeals where it is necessary to adopt an empirical outlook to accept the basis of many procedures that are understood as rational. As a single instance, the Bernoulli–Euler theory of bending does not apply immediately or rationally to the analysis of a reinforced concrete section because of the concentration of tensile strains at the cracks on the tension side of the neutral axis. However, the application of this theory is justifiable on the basis that it has proved to be a suitable predictor of the strength of reinforced concrete sections. The appeal to experience embedded in this statement is known as an empiricist appeal. In an empiricist appeal, a challenge to the rationality of a design procedure is rebutted on an empirical basis. Many similar instances can be found in reviewing the steel, wood, masonry and concrete codes. In many situations, engineers choose to use empirically determined minimum values of concrete reinforcement, steel sizes, masonry sizes or other decisions, without necessarily consulting the codes or rationally based formulas.

There is also utility for an empirical outlook in the present use of data-based models for the application of artificial intelligence to solve engineering problems. In these methods, inputs and outputs are compared over a very large dataset and a computer is trained to produce matching outputs simply based on experience, rather than considering any cause.

The greatest current usefulness of empirical design is the ability to estimate quickly the sizes, materials and configuration of a proposed design. When these important values can be determined empirically, there is a starting point for a rational design or a basis for checking the results of rational design. Empirical design in this form consists of span:depth, height:thickness and other ratios, and some knowledge of the minimum feasible sizes of a proposed building. This ability to determine suitable initial values of the size of a structural element is a very useful adjunct to the design methods taught to engineers as part of their training. The ability to determine the appropriate size with a minimum of effort allows a student to correct their own design work when proceeding analytically, and further allows the overall configuration of a structure to be taught in addition to the detailed design of the elements of a structure.

Empirical design is also of value in specialised circumstances. In the analysis of historic structures, there is often an absence of design methods for a historic feature, such as cast iron compression members in a truss. In such cases, it may be justifiable to make an empiricist appeal that the structure is working whether or not we can analyse it. This is often a better response than determining that a historic structure requires strengthening because its analysis is uncertain. Empirical design is invoked in investigations of building envelopes, where it is necessary to follow the path of the water through the envelope empirically before it is justifiable to use building science to determine the cause of problems that a wall or roof may be experiencing.

Empirical design can be used in any routine design situation. The repeated design of a three-storey wood-framed apartment block does not require further detailed investigation when the design of a similar structure is under consideration. Speciality engineers who design four or five storey wood-framed apartment buildings may not need to modify the design decisions from project to project, as the module of the apartments is the same, the specified materials are the same, the heights are the same and so on. So, the stud schedule for bearing walls need not change from project to project. However, the skill of an experienced designer is required to recognise where an unusual situation would need some design attention.

Reinforced concrete foundations are a further example of a structure that can be designed with only minimal appeals to reason, with the bulk of the design of this feature by empirical methods. The situation for reinforced concrete one-way slabs is similar. A 110-mm-thick reinforced concrete one-way slab can span approximately 16 times its thickness, or 1800 mm and minimum reinforcement of 0.2% reinforcement will be sufficient. These are examples of empirical information that practising engineers may choose to use for design, in spite of the availability of analytical methods. These sorts of empirical procedures have enduring value in structural engineering.

So, according to this discussion of empirical engineering and design, we can recognise that empirical design is an effective and enduring means of determining the size, configuration, material and details for the elements of a building structure, and that an understanding of empirical design has a place in contemporary engineering education and practice.

Boothby T
ISBN 978-0-7277-6633-5
https://doi.org/10.1680/edse.66335.103

Index

Page numbers for figures are in *italics*, tables are in **bold**.

Printed and bound by CPI Group (UK) Ltd, Croydon, CR0 4YY

02/11/2023

08160054-0001